Oxford Chemistry Series

General Editors

P. W. ATKINS   J. S. E. HOLKER   A. K. HOLLIDAY

Oxford Chemistry Series

R. J. Puddephatt
DEPARTMENT OF CHEMISTRY, UNIVERSITY OF WESTERN ONTARIO, CANADA
and
P. K. Monaghan
SIR WILFRED GRENFELL COLLEGE, MEMORIAL UNIVERSITY OF
NEWFOUNDLAND, CANADA

The Periodic Table of the Elements

Second Edition

Clarendon Press · Oxford · 1986

Oxford University Press, Walton Street, Oxford OX2 6DP
Oxford   New York   Toronto
Delhi   Bombay   Calcutta   Madras   Karachi
Kuala Lumpur   Singapore   Hong Kong   Tokyo
Nairobi   Dar es Salaam   Cape Town
Melbourne   Auckland
and associated companies in
Beirut   Berlin   Ibadan   Nicosia

Oxford is a trade mark of Oxford University Press

Published in the United States
by Oxford University Press, New York

British Library Cataloguing in Publication Data
Puddephatt, R. J.
The periodic table of the elements.—2nd ed.
—(Oxford Chemistry series; 32)
1 Periodic law
I. Title II. Monaghan, P. K.
541'.9'01   QD467
ISBN 0-19-855515-6
ISBN 0-19-855516-4

Library of Congress Cataloging in Publication Data
Puddephatt, Richard J.
The periodic table of the elements.
(Oxford chemistry series; 32)
Bibliography: p.
Includes index.
1. Periodic law. I. Monaghan, P. K. II. Title.
III. Series.
QD467.P83   1985   541'.9'01   85-29138
ISBN 0-19-855515-6
ISBN 0-19-855516-4 (pbk.)

Set by Spire Print Services Ltd., Salisbury, Wilts
Printed in Great Britain by St Edmundsbury Press
Bury St Edmunds, Suffolk

Editor's Foreword

The periodic table of the chemical elements has existed for more than a century. During that time it has provided a basis, and also a stimulus, for enormous developments in inorganic chemistry, and it continues to play an important part in the further extension of the subject. Increasingly, however, it must also provide a framework which students can use to gain some knowledge and understanding of the vast chemistry of more than one hundred elements and their compounds. This book is designed to provide, without a detailed discussion of theoretical concepts, an account of the way in which the periodic table is constructed, and of the trends in properties, both physical and chemical, which are revealed by the table.

A.K.H.

Acknowledgments

The authors wish to thank Professor A. K. Holliday, who suggested that this book should be written and suggested several improvements to the text. They are also grateful to Professor C. E. H. Bawn for his encouragement, and to the Clarendon Press for many helpful comments. Thanks are also due to the authors and publishers who permitted reproduction of diagrams; specific acknowledgments are given in the text. Finally, R.J.P. would like to thank his wife for her assistance at various stages of the book's preparation.

Preface to the Second Edition

The main purpose of this second edition remains the same as the first. It is to present, at a level suitable for first year students in universities and technical colleges and for senior secondary school students, the periodic behaviour of the elements and their compounds. We have tried to emphasize the fact that the periodic table is a powerful, unifying tool in the hands of the chemist.

The book has been expanded to include a slightly longer section on bonding, along with some additional factual information on the elements and their compounds. We have deliberately avoided making the book too theoretical or overburdening it with a large number of facts, so that the central theme will remain clear. Students wishing to expand their knowledge of chemical bonding or to learn more about the properties of the elements and their compounds are recommended to read the more comprehensive books given in the appendix.

London, Ontario   R. J. P.
March 1985   P. K. M.

Contents

1. Introduction

To achieve a thorough understanding of so complex a subject as chemistry, it would be highly desirable to fit all the facts into a simple, logical pattern. The periodic table of the elements, though less than ideal, has served to systematize the properties of the elements for well over 100 years. This book explores the basis of the observed periodicity of properties of the elements and their compounds, and describes many examples to illustrate these periodic trends.

The early history of ideas leading up to the periodic table is a fascinating subject but will not be treated here. The concept came to fruition in the years 1860–1870. By 1860 about 60 elements were known and, in the same year, a consistent set of atomic weights was presented by Cannizzaro. Two years later Newlands reported that, when the elements are arranged in order of increasing atomic weight, the properties of the eighth element resembled those of the first, those of the ninth were similar to those of the second, and so on. The noble gases had not yet been discovered. This relationship he called the 'law of octaves'. He presented his ideas to the London Chemical Society in 1864. The Society refused to publish his work.

The first periodic table of the elements was drawn up by the Russian chemist Mendeleev in the years 1868–1870; one version of Mendeleev's periodic table is shown in Fig. 1. It was obtained by arranging the chemical elements in order of increasing atomic weights. Vertical columns of elements were called *groups*, horizontal rows of elements, *periods*. Elements in the same vertical group possessed similar chemical properties.

This discovery of the periodic behaviour of the chemical elements provided an important stimulus for further chemical investigations. In the time of Mendeleev there were many 'holes' in the periodic table, where elements had not yet been discovered. Mendeleev was able to predict the chemical properties of each of these elements on the basis of the position in the periodic table. His predictions were found to be amazingly accurate when the elements were later discovered and their chemical properties investigated.

However it soon became apparent that a periodic table based on the atomic weights of the elements was not entirely satisfactory. For instance it was shown that tellurium had a higher atomic weight than iodine despite the fact that the opposite arrangement in the periodic table was clearly required by their chemical properties.

It was not until the electronic structure of atoms had been established that the full significance of the periodic table could be recognized. A brief account of our present knowledge about atomic structure is given in Chapter 2, and the way in which the modern periodic table is built up logically on the basis of this knowledge is discussed in Chapter 3.

It is now clearly established that the periodic table is based on the electronic structures of the element atoms and thus the elements are arranged according to the number of electrons or protons that they possess. In its modern expression the Periodic Law states that 'the properties of the chemical elements are a periodic function of the atomic number'. The importance of the periodic table in chemistry has steadily grown and two aspects in particular deserve special mention.

First of all, there has been a great increase in recent years in our knowledge of the physical properties of the chemical elements and their compounds, largely due to the use of new structural methods. These properties can often be readily understood in terms of the electronic structures of the elements and their positions in the periodic table. In turn these physical properties provide a basis on which the chemical properties of the elements and their compounds can be rationalized. Periodic trends in the physical properties of the elements are discussed in Chapters 4 and 5 while subsequent chapters are concerned with periodic trends in chemical properties of the elements and their compounds.

Group I	Group II	Group III	Group IV	Group V	Group VI	Group VII	Group VIII
–	–	–	RH_4	RH_3	RH_2	RH	–
R_2O	RO	R_2O_3	RO_2	R_2O_5	RO_3	R_2O_7	RO_4
$H=1$							
$Li=7$	$Be=9$	$B=11$	$C=12$	$N=14$	$O=16$	$F=19$	
$Na=23$	$Mg=24$	$Al=27$	$Si=28$	$P=31$	$S=32$	$Cl=35\cdot5$	
$K=39$	$Ca=40$	$?=44$	$Ti=48$	$V=51$	$Cr=52$	$Mn=55$	$\{$ $Fe=56$ $Co=59$ $Ni=59$
$Cu=63$	$Zn=65$	$?=68$	$?=72$	$As=75$	$Se=78$	$Br=80$	
$Rb=85$	$Sr=87$	$Yt=88$	$Zr=90$	$Nb=94$	$Mo=96$	$?=100$	$\}$ $Ru=104$ $Rh=104$ $Pd=106$
$Ag=108$	$Cd=112$	$In=113$	$Sn=118$	$Sb=122$	$Te=128$	$I=127$	
$Cs=133$	$Ba=137$	$Di=138$	$Ce=140$				
		$Er=178$	$La=180$	$Ta=182$	$W=184$		$\{$ $Os=195$ $Ir=197$ $Pt=198$
$Au=199$	$Hg=200$	$Tl=204$	$Pb=207$	$Bi=208$			
			$Th=231$		$U=240$		

FIG. 1. Mendeleev's periodic table. The notation used indicates the atomic weights of the elements, or the predicted values for elements which were still to be discovered; for instance, ? = 68 for the element now known as gallium.

The second important aspect of the periodic table is seen in its almost universal application in inorganic chemistry. The periodic table has played a major part in the great resurgence of interest in inorganic chemistry; the discovery of new elements and compounds has been stimulated on countless occasions by a study of the periodic table and its possible amplification and extension. Moreover, the vast expansion of knowledge about the elements and their compounds can only be comprehended in terms of the periodic table; it is no longer possible, or even desirable, to learn the properties of each chemical element or compound in isolation. Because broad trends in chemical behaviour can readily be rationalized using the periodic table, inorganic chemistry need no longer be regarded as a mere catalogue of facts. In this short book it is hoped to show, by concentrating on *comparisons* between and *trends* in the chemical properties of the elements in terms of their positions in the periodic table, rather than on individual compounds, that the study of inorganic chemistry can be a fascinating and stimulating subject.

2. Atomic structure

The modern periodic classification of the elements depends upon atomic structure, and in particular upon the way in which the energy levels of electrons in atoms are arranged. These energy levels are defined in terms of whole numbers, called *quantum numbers*, and some understanding of their origin is necessary before the detailed structure of the periodic table is discussed further.

The hydrogen atom with one electron is the simplest possible atom and has served as a model for all modern theories of atomic structure. Niels Bohr in 1913 proposed the first theory of the hydrogen atom, making use of the quantum theory put forward by Planck a few years before.

Bohr theory of the hydrogen atom

The Bohr model of the hydrogen atom may be pictured as the electron moving in a circular orbit about the proton like the earth circling the sun (Fig. 2).

The electrostatic attractive force between the electron and the proton is balanced by the centrifugal force on the electron due to its circular motion. This leads naturally to the equation

$$\frac{1}{4\pi\varepsilon_0} \cdot \frac{e^2}{r^2} = \frac{m_e v^2}{r} \tag{1}$$

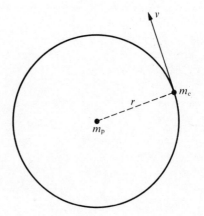

FIG. 2. The Bohr model of the hydrogen atom.

where e, m_e, and v are respectively the charge, mass, and velocity of the electron and r is the radius of the orbit as defined in Fig. 2. ε_0 is a constant known as the permittivity of free space.

So far the picture is the same as one proposed earlier by Rutherford in which the orbital radius r could take any value. Bohr now made the apparently startling assumption that the angular momentum of the electron, $m_e vr$, was quantized, that is that it could take only certain discrete values. So

$$m_e vr = n \cdot \frac{h}{2\pi},$$

where h is a fundamental constant called Planck's constant and n is an integer ($n = 1, 2, 3, \ldots$). Substituting for v in equation 1 we obtain

$$\frac{1}{4\pi\varepsilon_0} \cdot \frac{e^2}{r^2} = \frac{m_e}{r} \cdot \frac{n^2 h^2}{4\pi^v m_e^2 r^2},$$

which on rearrangement gives

$$r = \frac{n^2 h^2 \varepsilon_0}{\pi m_e e^2}.$$

This equation then gives the allowed radii of the electron orbits in the hydrogen atom in terms of the *quantum number, n*. For each allowed radius r_n there is an allowed energy of the electron E_n, i.e. $E_1, E_2, \ldots$, which can easily be calculated.

Bohr's theory was most successful in rationalizing known facts concerning the hydrogen atom. However the theory was not so successful when applied to atoms with more than one electron and it was soon modified and then discarded.

One of the chief objections to the Bohr theory was the rather arbitrary way in which quantum numbers were introduced. Another difficulty was that the theory assumed the electron to behave only as a particle. Future theories were to take account of the wave properties of the electron from which the quantum numbers evolve naturally.

The wave theory of electrons
At the time when Bohr proposed his theory of the hydrogen atom it was already known that the photon, the smallest unit of light, showed both particle and wave properties. The energy of the photon is given by the Planck–Einstein expressions

$$E = h\nu = mc^2,$$

where h is Planck's constant, ν the frequency of the light wave, m the mass of the photon, and c the velocity of light.

FIG. 3. Representation of an electron wave. The solid line represents a possible stationary wave, while the dotted line shows how a wave of somewhat different wavelength would be destroyed by interference.

Now the momentum p of the photon is given by

$$p = mc = \frac{h\nu}{c}.$$

Since the wavelength λ and frequency ν of light are related by the expression $\lambda = c/\nu$, we have

$$\lambda = \frac{c}{\nu} = \frac{h}{p}.$$

De Broglie suggested that this expression, which gives the wavelength associated with a particle having momentum p, should hold for all matter, and in particular that it should hold for a electron in a 'Bohr orbit'. In such a situation it is necessary to have a standing wave, i.e. there must be an integral number of waves in Bohr orbit (since otherwise the wave would be destroyed by interference, as shown in Fig. 3). Thus we get the expression

$$2\pi r = n\lambda = n \cdot \frac{h}{p},$$

which on rearrangement gives the following expression for the angular momentum rp of the electron in a Bohr orbit:

$$rp = n \cdot \frac{h}{2\pi}.$$

Thus the quantization of angular momentum which was assumed arbitrarily by Bohr follows naturally from this wave treatment. It also follows that the electronic energy levels must be quantized.

The Schrödinger equation

Since an electron has wave properties, its behaviour can be described in terms of a wave function ψ. The physical significance of ψ is not easily grasped but it is best thought of as analogous to the amplitude of the wave. The probability of finding an electron in any volume is proportional to the integral of the square of the absolute value of the wave function over that volume, or probability $(x, y, z) \propto |\psi(x, y, z)|^2$.

The relationship between the energy of the electron in the hydrogen atom and its wave function was first given by Schrödinger. This relationship forms the basis of modern quantum mechanics and is known as the Schrödinger equation.

$$H\psi = E\psi$$

where H is a mathematical operator called the Hamiltonian operator and represents the general form of the kinetic and potential energies of the system and E is the numerical value of the energy for a given wave function ψ.

The Schrödinger equation can be solved for certain wave functions ψ and these are known as *eigenfunctions*. The corresponding allowed energy levels are called *eigenvalues*.

$$\psi_{n,l,m}(r, \theta, \phi) = R_{n,l}(r)\Phi_{l,m}(\theta, \phi).$$

If written in terms of polar co-ordinates r, θ, ϕ the wave equation can be split up into a radial component $R_{n,l}(r)$ and an angular component $\Phi_{l,m}(\theta, \phi)$. The square of the radial part of the wave function $|R_{n,l}(r)|^2$ gives the probability of finding the electron at a distance r from the nucleus. The wave function can only be an eigenfunction when n is an integer; n is called the principal quantum number and defines the mean separation of the electron from the nucleus. Similarly the wave function can only be an eigenfunction when l is an integer. l is known as the azimuthal quantum number and defines the angular momentum of the electron; it can take the values $l = 0, 1, 2$, up to $(n - 1)$.

The angular part of the wave function defines a third quantum number

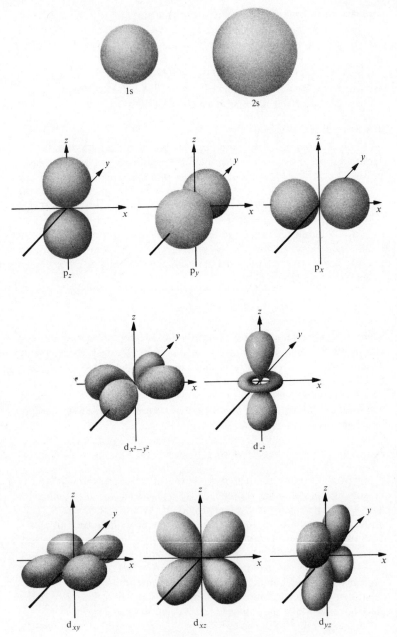

FIG. 4. Representations of atomic orbitals.

m called the magnetic quantum number. This gives the allowed values of the component of the electronic angular momentum in a given direction when a magnetic field is applied and can take all integral values between l and $-l$.

Atomic orbitals

The hydrogen wave functions $\psi_{n,l,m}$ are known as *orbitals*. The orbitals are broadly classified according to the value of l. When $l = 0$ we have an s orbital which is spherically symmetrical, when $l = 1$ were have a p orbital, when $l = 2$ a d orbital, and when $l = 3$ an f orbital. The orbitals are then prefixed by the value of n, so that for instance if $n = 2, l = 1$ we have a 2p orbital.

The p orbital can have various orientations in space depending on the value of m which can be 1, 0, or -1. Similarly there are *five* possible orientations of d orbitals, as $m = 2, 1, 0, -1,$ or -2, and *seven* possible orientations of f orbitals.

The orbitals may be represented by drawings showing the regions of space in which there is a high probability of finding the electron[†]. Examples of these representations of s, p, and d orbitals are given in Fig. 4. They show very clearly the directional character of the p and d orbitals.

The spin quantum number

Just as the earth spins about its own axis while circling the sun, so may the electron be regarded as spinning about its own axis. The angular momentum due to this spin is quantized so that its component in the direction of a magnetic field is given by s. $h/2\pi$, where s is the spin quantum number and can take only values $+\frac{1}{2}$ or $-\frac{1}{2}$.

The spin quantum number was introduced to explain some experimental results, and is not obvious in the Schrödinger equation. However, a more sophisticated theory of quantum mechanics, Dirac's relativisitic treatment, leads naturally to this fourth quantum number.

Overall then the state of the electron in a hydrogen atom is defined by *four* quantum numbers. Three of these, $n, l,$ and m, are associated with the motion of the electron about the proton while the fourth, s, is associated with the spin of the electron about is own axis.

PROBLEM

2.1. Write down the possible quantum numbers $n, l, m,$ and s for all (a) 2s (b) 2p (c) 3d orbitals. If no two electrons can have all four quantum numbers the same, what is the maximum number of electrons that can be accommodated in these orbitals?

[†] These drawings represent only the directional character of the orbitals and are not intended to represent the actual shapes of the orbitals.

3. Building up the periodic table

So far we have shown that the energy levels of an electron in a hydrogen atom are defined by four quantum numbers n, l, m, and s. The electron will normally be in its lowest energy or *ground* state; this state is the state with $n = 1$ and $l = 0$ so that in the ground state the electron will be in the 1s orbital.

For more complex atoms we can still define the electronic energy levels in terms of the four quantum numbers, and we shall therefore want to know the values of these quantum numbers for all the electrons in the atom. There is an important principle first defined by Pauli which determines the permitted quantum numbers for the electrons in an atom.

The Pauli Exclusion Principle

The Exclusion Principle states that no two electrons in the same atom can have all four quantum numbers the same. Thus in the helium atom, which has two protons and two electrons, both electrons can enter the 1s orbital; one will have the quantum numbers $n = 1$, $l = 0$, $s = +\frac{1}{2}$, and the other $n = 1, l = 0, m = 0, s = -\frac{1}{2}$. According to the exclusion principle the 1s *electron shell* is now filled. In the lithium atom, which has three electrons, the first two electrons fill the 1s orbital so that the third electron must be accommodated in an orbital with the principle quantum number $n = 2$. When $n = 2, l$ can take the values 0 or 1 so that the next electron could enter either the 2s or the 2p orbital.

Stability of atomic orbitals

We have already seen that the 1s orbital is more stable than the 2s orbital because the electron in the 1s orbital is on average closer to the nucleus and so feels a greater electrostatic attraction to the nucleus than the 2s electron. For the hydrogen atom the 2s and 2p orbitals have identical energies but this is not the case for large atoms.

Let us look at the lithium atom in more detail. The two 1s electrons are directly attracted by the nuclear charge of the three protons and are very firmly held. An electron in the outer orbital with $n = 2$ has its greatest density outside the 1s shell and so is *shielded* from the nucleus by the two 1s electrons. It therefore 'feels' an *effective nuclear charge* of only a single proton and is much more weakly held. This is a somewhat oversimplified picture; the $n = 2$ electron cloud *penetrates* to some extent within the boundary of the 1s orbital, and this component is naturally more strongly held to the nucleus.

Now the 2s orbital penetrates the 1s to a greater extent than does the 2p (this can be seen in a qualitative way by inspection of the representations of orbitals in Fig. 4, p. 8) and so the 2s orbital is more stable than the 2p in the lithium atom. Thus the *electron configuration*, that is the way in which the electrons are distributed among the atomic orbitals, for the lithium atom in its ground state is $1s^2 2s^1$, the occupation numbers of the orbitals being written as superscripts.

In general the stability of orbitals in multi-electron atoms can be summarized as follows: if l is constant the orbital with lower principal quantum number is more stable, so we have the stability series for example $1s > 2s > 3s > 4s$. Similarly if n is constant the orbital with lower l is more stable, so the stability series $4s > 4p > 4d > 4f$ arises.

The periodic table

We have now established the principle by which the periodic table is built up. Each new element in the table contains one more nuclear charge than the preceding element: this charge is neutralized by addition of an electron which is fed into the lowest energy orbital available, taking account of the Pauli Exclusion Principle. The elements can then *be classified naturally into periods depending on which electron shell is being filled*. Each period begins with the occupation of an *n*s orbital and ends when the *n*p orbitals are full.

A modern version of the periodic table is shown in Fig. 5, In the following sections we shall discuss its structure in some detail.

The short periods

The first period contains only the elements hydrogen and helium. Of all the element atoms helium has the least ability to take part in chemical combination, having very little tendency either to gain or to lose an electron (see ionization energies in Fig. 7, p. 18). For this reason helium is known as a *noble* (chemically unreactive) gas. We shall see that all subsequent periods are also completed by a noble gas.

The next period comprises the eight elements from lithium to neon; it is generally known as the *first short period*, or simply the *first row*. In this period the 2s and then the three 2p orbitals are consecutively occupied. In atoms having partially filled 2p orbitals the electrons prefer to maintain their spins mutually parallel, that is to have the same spin quantum numbers whenever possible. For this reason nitrogen has the electron configuration $1s^2 2s^2 2p_x^1 2p_y^1 2p_z^1$, which can conveniently be represented as as

$$\text{⑪} \quad \text{⑪} \quad \text{①①①}$$
$$1s \quad 2s \quad 2p$$

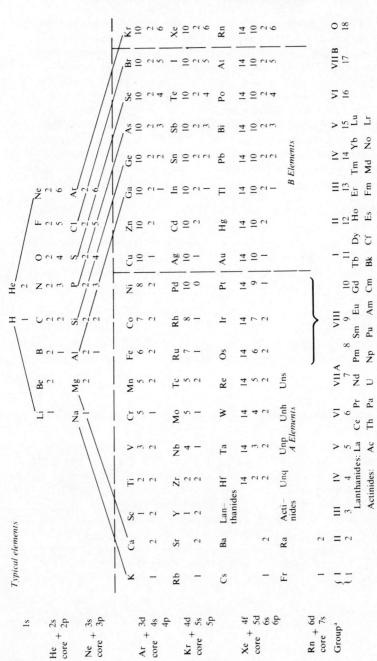

FIG. 5. A modern periodic table of the elements (after Phillips and Williams, 1965).

[a]The new IUPAC system is given below the traditional system.

where each circle represents an orbital and each arrow an electron, with the direction of the arrow indicating the orientation of the electron spin.

This electron configuration, rather than other possibilities having paired spins in the 2p orbitals, arises as a result of Hund's rule of maximum multiplicity. The rule simply states that the ground state of an atom will be that having the greatest multiplicity or the maximum value for the spin.

The final element in this period is neon, which has the electron configuration $1s^22s^22p^6$ and the orbitals with principal quantum number $n = 2$ are then completely occupied.

During the second short period, which comprises the elements from sodium to argon, the 3s and 3p orbitals are occupied in an exactly analogous manner to that just described and the final element, argon, therefore has the electron configuration $1s^22s^22p^63s^23p^6$. One might expect the 3d orbitals to the occupied next, but these orbitals do not penetrate the *electron core*† to any great extent and are much higher in energy than the 3s and 3p orbitals at this point. In fact, at argon the 4s orbital is lower in energy than the 3d orbitals and it is the 4s orbital which is occupied next to start the first long period.

The long periods

As predicted the next period begins with the filling of the 4s orbital, but then an important difference from the sequence of the short periods arises since the 3d orbitals are occupied before the 4p. The ten elements beginning with scandium make up the first *transition series* of elements as the five 3d orbitals are filled. The first long period is completed by the six elements from gallium to krypton as the 4p orbitals are occupied, so that altogether there are eighteen elements in the period.

We might expect these transition elements to have the electron configuration (Ar core) $3d^x4s^2$ where $x = 0$ to 10. This is generally the case although, exceptionally, chromium has the ground state configuration (Ar core) $3d^54s^1$ and copper (Ar core) $3d^{10}4s^1$. These exceptions indicate the added stability of filled or half-filled shells of d electrons, and also show how close in energy are the 3d and 4s orbitals. We shall see later that the 3d electrons are stabilized with respect to the 4s on crossing the transition series, so that by the end of the series the 3d orbitals have entered the electron core and are no longer involved in chemical bonding.

The structure of the second long period is analogous to that of the first, as the consecutive occupation of the 5s, 4d and 5p orbitals gives the

---

†The *electron core* comprises all electrons in an atom which are so tightly bound to the nucleus that they take no part in chemical combination. At argon all of the electrons are core electrons and the configuration $1s^22s^22p^63s^23p^6$ will in future be written in shorthand form as (Ar core).

eighteen elements from rubidium to xenon. Again in the transition series there are some unusual electron configurations as a result of the 5s and 4d orbitals having very similar energies for these elements. Elements not having the $4d^x5s^2$ configuration are niobium, $4d^45s^1$ molybdenum, $4d^55s^1$; ruthenium, $4d^75s^1$; rhodium, $4d^85s^1$; palladium, $4d^{10}5s^0$; and silver, $4d^{10}5s^1$. We see that six out of the ten elements from yttrium to cadmium are 'anomalous' in this series.

The third long period is even more complex, and the following section should be studied along with the periodic table given in Fig. 5, p. 12. It begins with the occupation of the 6s orbital giving the elements caesium and barium. Lanthanum is the next element; it has the ground state electron configuration (Xe core) $5d^16s^2$, and we might expect the third transition series to follow. However the seven 4f orbitals are occupied next, giving rise to a series of fourteen elements known as the *lanthanides*. It is not until the 4f orbitals are completely filled that further electrons enter the 5d orbitals, thus completing the third transition series; the period is completed with the occupation of the 6p orbitals giving the six elements from thallium to radon. This is the longest period and contains thirty-two elements in all.

The final period begins with the elements francium, radium, and actinium which have the electron configurations (Rn core) $7s^1$, $7s^2$, and $6d^17s^2$ respectively. The 5f orbitals are then filled giving a further series of fourteen elements from thorium to lawrencium known as the *actinide* elements. Further elements would be expected to form a fourth transition series but many have yet to be discovered. All the elements of this period are radioactive; many are very short-lived and have been synthesized in only minute quantities. For these reasons little is known about the chemistry of several of the elements.

Stability of atomic orbitals in the long periods

We saw that during the first transition series the 3d orbitals slowly became more stable than the 4s. A related effect is observed during subsequent long periods. Thus on crossing the lanthanide series the 4f orbitals are stabilized with respect to the 5d and 6s orbitals, so that by the end of the series they have entered the electron core. The overall variation of the stability of atomic orbitals with atomic number (the total number of electrons in an atom) is shown in Fig. 6. It shows clearly how the d and f orbitals change rapidly during the transition series from being high-energy orbitals to stable core orbitals

Classification into groups

The classification of the elements into groups is shown in Fig. 5. In the short periods the elements are conveniently classified into groups I–VII

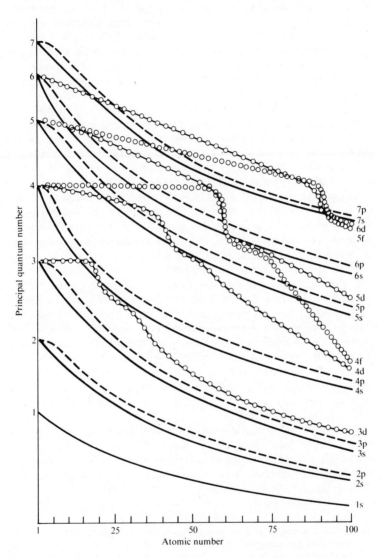

FIG. 6. The energies of atomic orbitals as a function of atomic number (after Cotton and Wilkinson, 1962).

with the noble gases as group 0[†]. As we shall see later (p. 23) the group number is, with some exceptions, the same as the maximum oxidation state[‡] shown by the elements of the group. Elements within a group generally show strong similarities in their chemical behaviour.

In the long periods, a further classification of the elements as A or B elements is made but, unfortunately, the conventions used in Britain and in North America are different. The need for the A and B nomenclature arises because the elements titanium and germanium, for example, each have a maximum oxidation state of four but clearly fall in different groups. In Britain, titanium is placed in group IVA and germanium in group IVB, but the opposite nomenclature is used in North America. The British nomenclature is used in Fig. 5, with the pre-transition metal elements and early transition elements labelled A elements and the late transition elements and post-transition elements labelled B elements.

As can be seen from Fig. 5, the elements iron, cobalt, and nickel do not fall naturally into either classification. These elements and their analogues in subsequent periods are placed in group VIII after the first classification by Mendeleev, though few of the elements ever achieve an oxidation state of eight and differences in chemistry between the elements iron, cobalt, and nickel are greater than the similarities.

It is probable that the new system of classification suggested by the International Union of Pure and Applied Chemistry will be adopted in the future. In this system the main group elements and transition elements are placed in groups 1–18 according to the total number of valence shell electrons for the elements in the long periods. This avoids the confusion of the A and B elements of the old system. As can be seen in Fig. 5, Group IVA now becomes Group 4 but Group IVB becomes Group 14. The anomalous Group VIII is separated into Groups 8, 9, and 10 in the proposed new system. For the next few years, the student will probably need to learn both systems.

The elements may also be classified more directly according to the orbitals being filled. Thus the elements of groups IA and IIA are collectively called the s-block elements, those of groups IIIB–VIIB are the p-block, the transition metals are the d-block, and the lanthanides and actinides are the f-block elements.

PROBLEM

3.1. Predict the ground-state electron configurations of elements having the atomic numbers (the number of protons in the nucleus) 4, 16, and 22. Identify the elements and classify them according to (a) the group and (b) the block of the periodic table into which they fall.

[†] Strictly this group should be called group VIII but group VIII was used by Mendeleev to classify some transition elements (see p. 2).
[‡] The term 'oxidation state' is discussed on p. 31.

4. Periodicity and the atomic and bonding properties of the main-group elements

Having established the periodic table of the elements, the remainder of this book is devoted to developing an understanding of the physical and chemical properties of the elements in terms of their positions in the periodic table. In this chapter we shall discuss trends in certain physical properties of the elements which can be used as a basis for discussing chemical behaviour.

Ionization energies of the elements

The *ionization energy* of an atom is the minimum energy required to remove an electron from a gaseous atom in its ground-state electron configuration. So for the magnesium atom the first ionization energy[†] is the energy required for the process

$$Mg(g) \rightarrow Mg^+(g) + e^- \qquad I = 7.4 \times 10^5 \text{ J mol}^{-1}$$

Similarly the second ionization energy is the energy required to remove a second electron

$$Mg^+(g) \rightarrow Mg^{2+}(g) + e^- \qquad I = 14.5 \times 10^5 \text{ J mol}^{-1}.$$

As might be expected, it is more difficult to remove an electron from a positively charged ion (cation) than from a neutral atom.

The first ionization energies of the elements of the short periods are plotted in Fig. 7. It is immediately apparent that the first ionization energy for lithium is much lower than that for helium. This is because the single 2s electron in lithium 'feels' a low effective nuclear charge since it is strongly shielded by the helium core $1s^2$ electrons; it is therefore easily removed. There is another drop in ionization energy between beryllium and boron. The low ionization energy for boron can be explained in terms of its electron configuration (He core) $2s^2 2p^1$; the 2p electron is shielded from the nucleus by the electrons in the filled 2s orbital and is readily lost. In subsequent elements the electrons in the 2p orbitals shield each other rather ineffectively from the nuclear charge so that the ionization energy increases as the 2p orbitals are filled[‡], until a maximum value is reached at

[†]The ionization energy, I, is usually defined for one mole of atoms and is measured in Joules.
[‡]Fig. 7 shows that the increase is not regular; there is a drop after nitrogen, which has the half-filled shell configuration $1s^2\, 2s^2\, 2p^3$.

Fig. 7. Ionization energies for the elements of the short periods.

the noble gas neon. There is another break at the first element of the next period, sodium, in which the single 3s electron is effectively shielded by the neon core electrons.

In general then, the ionization energies increase irregularly on crossing a period.

Comparison of the ionization energies within a group is also instructive, as shown in Fig. 8, which contains a plot of the ionization energies for the elements of groups IA and IB. The steady decrease in ionization energies in group IA (lithium to caesium) is due to the increasing *size* of the atoms on descending the group. The single 6s electron in the large caesium atom is much more readily removed than the 2s electron in lithium since the

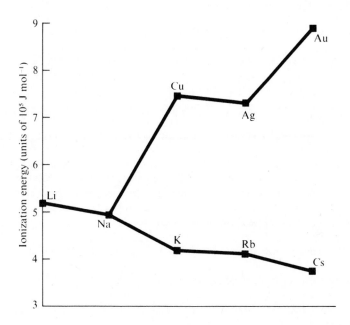

FIG. 8. Ionization energies for the group IA and IB elements.

electrostatic attraction of the electron to the nucleus decreases with distance.

For the group IB elements the ionization energies are considerably higher, and there is no general decrease on descending the group. The ionization process for copper is

$$Cu[(Ar\ core)\ 3d^{10}4s^1] \rightarrow Cu^+[(Ar\ core)\ 3d^{10}] + e^-.$$

The 4s electron is sheilded by the argon core electrons and the filled subshell of d electrons. Now the strongly directional and diffuse d electrons shield the 4s electron rather ineffectively from the nuclear charge; it therefore 'feels' a high effective nuclear charge and the ionization energy is much higher than that of potassium, which has the configuration (Ar core) $4s^1$.

The ionization energy for the larger silver atom is, as expected, somewhat less than that for copper, but there is a large increase for gold. The ionization reaction for gold is

$$\text{(Xe core) } 4f^{14}5d^{10}6s^1 \rightarrow \text{(Xe core) } 4f^{14}5d^{10} + e^-.$$

The 6s electron is poorly shielded by both 4f and 5d electrons and a high ionization energy results. Relatively high ionization energies for the B elements and particularly for the post-lanthanide elements represent a general phenomenon which has important consequences for the chemistry of these elements.

Some chemical deductions can be made from ionization energy comparisons. Thus we expect that gaseous cations will be formed most readily by the group IA elements which have the lowest ionization energies, less readily by the group IB elements, and least readily by the group O noble gases. Second and subsequent ionization energies are also important. Thus Na^+ is more stable than Mg^+ with respect to their atoms; however, compare the ionization energies

$$Na^+ \text{(neon core)} \rightarrow Na^{2+} (1s^2 2s^2 2p^5), \quad I = 45.6 \times 10^5 \text{ J mol}^{-1};$$

$$Mg^+ \text{(neon core } 3s^1) \rightarrow Mg^{2+} \text{(neon core)}, \quad I = 14.5 \times 10^5 \text{ J mol}^{-1}.$$

The ion Mg^{2+} is important in chemical compounds but Na^{2+} is never formed.

Electron affinities of the elements

The electron affinity of an atom is the energy *released* when an electron is added to the atom in the gas phase, and is equal to the ionization energy of the negative ion

$$F(g) + e^- \rightarrow F(g)^- \qquad \text{E.A.} = 337 \text{ kJ mol}^{-1}$$

$$F(g)^- \rightarrow F(g) + e^- \qquad \text{I.} = 337 \text{ kJ mol}^{-1}.$$

When defined in this way, the electron affinities of most elements are positive but, when the atom has a closed shell configuration, addition of an extra electron may be unfavourable. For example, helium $(1s^2)$ and beryllium $(1s^2 2s^2)$ and their congeners[†] have negative electron affinities. Electron affinities are greatest for the halogen atoms which achieve the noble gas electron configuration on forming the uninegative ions. Some trends in electron affinities are given in Fig. 9.

[†]Congeners are members of the same group of the periodic table.

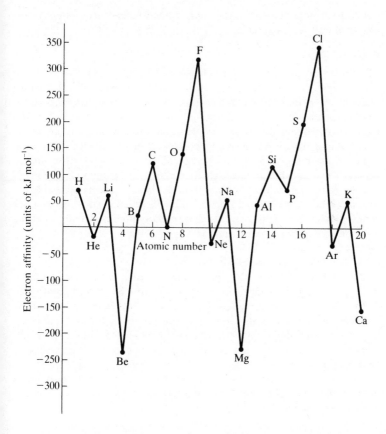

FIG. 9. First electron affinities of the elements in the two short periods.

The addition of another electron to a negatively charged ion is always unfavourable, even when this would enable the element to reach the noble gas electron configuration. This is due to the unfavourable electrostatic repulsion between the electron and the negatively charged ion. Consider, for example, the oxygen atom for which O^{2-} has the noble gas electron configuration

$$O(g) + e^- \rightarrow O^-(g) \qquad \text{E.A.} = 141 \text{ kJ mol}^{-1}$$

$$O^-(g) + e^- \rightarrow O^{2-}(g) \qquad \text{E.A.} = -780 \text{ kJ mol}^{-1}.$$

Clearly formation of $O^{2-}(g)$ (or any other doubly charged anion) from the element is unfavourable.

Chemical bonding

The theory of the chemical bond has reached a very sophisticated level. The discussion presented here is intended to give only the barest introduction to the nature of the chemical bond, and the reader is referred to specialized texts for a more rigorous treatment.[†] There are two extremes of bonding, ionic and covalent, but there is no sharp boundary between the two; both can be important in either molecules or solid lattices. In forming chemical bonds, the main group elements often seek to achieve the outer shell (valence shell) electron configuration of the noble gas closest to them in the periodic table. For most elements this noble gas electron configuration would be ns^2np^6 (with filled d and f orbitals for some heavier elements), but hydrogen and lithium would seek the electron configuration of helium ($1s^2$).

Ionic bonding

We shall consider the bonding in caesium chloride, CsCl. Caesium has a valence shell configuration of $6s^1$, whilst chlorine has a configuration of $3s^23p^5$. In the reaction between these two elements in the atomic states, caesium transfers the 6s electron to chlorine producing the Cs^+ cation and Cl^- anion with the stable electron configurations $5s^25p^6$ and $3s^23p^6$, respectively. An ionic bond is then formed between these oppositely charged ions.

The lowest ionization energy of any element (that for caesium) is greater than the highest electron affinity (that for chlorine), so that even the atomic reaction

$$Cs + Cl \rightarrow Cs^+ + Cl^-$$

would not take place if isolated ions were formed. In fact ionic *compounds* exist as regular arrays of cations and anions in solid lattices. The ionic lattice for caesium chloride is shown in Fig. 10. Each caesium ion is surrounded by eight chloride ions[‡]; it is attracted to these ions electrostatically, but the electron shells produce a counter-repulsion. The sum over the whole lattic of the energy arising from these attractive and repulsive forces is called the *lattice energy*; it is this energy which is responsible for the stability of solid ionic compounds.

[†]For example: C. A. Coulson, revised by R. McWeeny; *The shape and structure of molecules* (Clarendon Press, Oxford, 1982).

[‡]We say that the 'coordination number' of each caesium ion is eight.

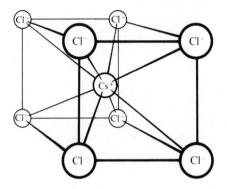

FIG. 10. Part of the lattice structure of caesium chloride.

For a lattice containing cations of charge Z^+ and anions of charge Z^- with a closest interionic separation of r, the lattice energy, U, is essentially proportional to Z^+Z^-/r. Hence lattice energies are greatest for small, highly charged cations and anions.

It is the balance between ionization energies and lattice energies that determines the stability of ionic compounds. For instance calcium forms the chloride $CaCl_2$ rather than $CaCl$ or $CaCl_3$. The extra lattice energy for the Ca^{2+} ion more than balances the extra energy needed to form this ion from Ca^+, but to remove a further electron, which would come from the argon electron core, would require a prohibitive amount of energy. It is easy to see why the ionic compounds of the group I–III elements usually contain the elements in their group valency states.

Generally speaking, formation of ionic bonds involves the loss or gain of no more than two electrons. Formation of triply charged anions or cations is far less common.

Covalent bonding

The first modern theory of the covalent bond considered that bond formation took place by the sharing of electrons between two atoms giving an electron-pair bond, for instance

$$H\cdot + \cdot Cl \rightarrow H : Cl.$$

Atoms of the main group elements can in this way achieve a filled valence shell.

In more recent (valence bond) theory, the electron-pair bond is considered to be formed by overlap of atomic orbitals on the two atoms.

Let us consider formation of methane from carbon and hydrogen atoms. The ground-state electron configuration of carbon is

$$⊕ ⊕ ①①○$$
$$\text{1s} \text{2s} \text{2p}$$

There are only two unpaired electrons so that only two covalent bonds can be formed. In order to form four bonds an electron must be 'promoted' from the 2s to the 2p orbital. The energy required is called the *promotion energy*. We now have the valence electron configuration at carbon

$$⊕ ① ①①①$$
$$\text{1s} \text{2s} \text{2p}$$

The 2s and three 2p orbitals of carbon are now considered to 'mix', forming four new, equivalent, *hybrid* orbitals known as sp^3−hybrid orbitals. The process described here is called *hybridization*. Each of the four sp^3−hybrid orbitals contains one electron and, by overlap with the 1s orbital of a hydrogen atom, forms a covalent bond as shown below.

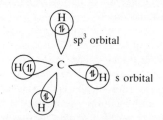

Formation of covalent bonds builds up electron density between the nuclei of the bonded atoms. The shapes of the overlapping orbitals determine the nature of the covalent bond. For example the H_2 molecule is formed by overlap of the two hydrogen 1s orbitals. The resultant bond is known as a covalent σ-bond.

In ethylene, $H_2C{=}CH_2$, the carbon atoms are doubly bonded, one of these bonds being a σ-bond. The second bond is formed by a sideways overlap of 2p orbitals on the carbon atoms, producing a covalent π-bond, in which there is a region of electron density above and below the plane of the molecule.

For the elements carbon, nitrogen and oxygen, both $\sigma-$ and $\pi-$bonding are important, but we shall see that π-bonding is less favourable for elements of the second or third row of the periodic table.

Electron-pair repulsion

The stereochemistry of methane (the orientation of the atoms in the molecule) is governed by the need to minimize electrostatic repulsions between the four bonding electron pairs. This is best achieved if the C—H bonds are directed towards the corners of a tetrahedron with the carbon atom at the centre and this is the observed stereochemistry of the methane molecule. We shall see that the stereochemistry of many simple covalent compounds can be deduced by similar considerations.

Covalent radii of the elements

Atomic size is a nebulous quantity since an atom, unlike such familiar objects as billiard balls, can have no well defined boundary. However, methods of estimating approximate atomic sizes have been developed and we shall consider trends in the covalent radii of the elements.

The nature and value of the concept of covalent radius is best illustrated by an example. Thus the single bond covalent radius for carbon is defined as half the C—C bond distance of 0.154 nm[†] in ethane, CH_3—CH_3, and that for fluorine as half the F—F distance of 0.142 nm in the fluorine molecule. We can then use these covalent radii to compute the expected C—F distance in methyl fluoride, CH_3—F, to be $(0.077 + 0.071) = 0.148$ nm. The observed and calculated bond distances usually agree well, though, for the above example, the observed C—F distance is in fact 0.138 nm. This contraction from the expected value is a result of the extra bond strength arising from the partial ionic

[†]1 nm = 10^{-9}m.

FIG. 11. Covalent radii of the elements of the short periods.

character of the C—F bond; by taking account of this factor, excellent agreement between calculated and observed covalent bond lengths may be obtained. The covalent radii may be used directly to calculate bond lengths when there is little or no ionic character in the bonds.

Periodic trends in covalent radii can be correlated with similar trends in ionization energies. Thus a high effective nuclear charge leads to a high ionization energy and to a low covalent radius as the outer electrons are tightly held and so are pulled in closer to the nucleus. Covalent radii for the elements of the short periods are given in Fig. 11. Covalent radii decrease on crossing a period, increase on descending a group and are greater for the A elements than for the B elements.

Ionic radii of the elements

We have seen how covalent radii can be estimated by examining bond lengths in covalent compounds. By a similar, though slightly more complex, process, ionic radii can be obtained from internuclear distances in ionic compounds.

Figure 12 shows the covalent radii for the group IVB elements together with the ionic radii for the M^{2+} and M^{4+} ions; it also shows the covalent

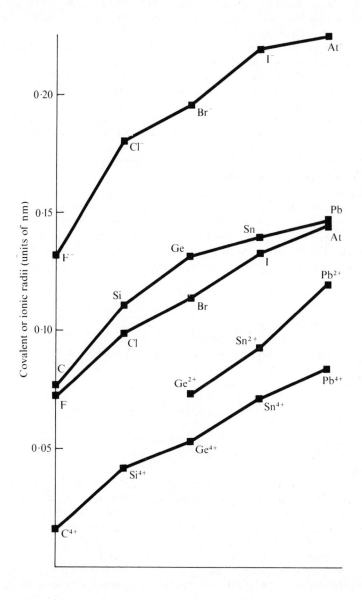

FIG. 12. Covalent radii and ionic radii for some element atoms or ions. (The values for C^{4+} and Si^{4+} are estimated, since no compounds containing these ions exist.)

radii of the halogens (group VIIB elements) and the ionic radii of the halide ions. The radii *decrease rapidly when the ions carry a positive charge* and *increase with negative charge*; they increase on descending a group. We can readily explain these trends in terms of the effective nuclear charge felt by the outer electrons.

For example, the ions K^+ and Cl^- are isoelectronic (both have the noble gas electron configuration of argon) but the ionic radii are K^+, 0.133 nm and Cl^-, 0.181 nm. The positively charged K^+ ion is smaller because the outer electrons feel a greater effective nuclear charge (K^+ nucleus has 19 protons) than the outer electrons of the Cl^- ion (with only 17 protons in the nucleus).

Electronegativities of the elements

The electronegativity of an element is a measure of the power of the element to attract electrons to itself *in chemical compounds*. It is thus different from the electron affinity which is a property of isolated atoms, as described on pages 28–30. We can illustrate this property by considering the hydrogen chloride molecule; the hydrogen–chlorine bond is intermediate in character between a covalent and an ionic bond:

$$H\text{-}Cl \qquad \text{and} \qquad H^+Cl^-.$$

(electron pair equally shared)   (bonding electrons all on chlorine)

Chlorine is the more electronegative element and so accumulates a net negative charge. We can represent hydrogen chloride as $H^{\delta+} - Cl^{\delta-}$ and write the overall wave function

$$\psi = \psi \text{ covalent} + c\psi \text{ ionic}$$

where c is a measure of the *ionic character* in the H-Cl bond and is determined by the difference in electronegativity between hydrogen and chlorine.

It is difficult to assign accurate electronegativities to the elements, partly because the precise values may vary in different compounds, but the concept is nevertheless extremely valuable. One quantitative estimate of electronegativity is the average of the ionization energy and the electron affinity for an element, that is the average of the ability of an atom of the element to donate or accept an electron.

Some electronegativity values for the elements of the first short period and for the group IA and IB elements are shown in Fig. 13. The trends are similar to those in the ionization energies shown in Figs. 7 and 8. Thus the electronegativities increase steadily on crossing a period and generally

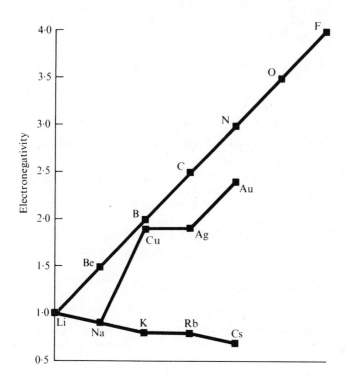

FIG. 13. Electronegativities of some elements.

decrease on descending a group, though the latter trend may be reversed for the B elements.

Various measures of electronegativity have been proposed and all give roughly parallel scales. Figure 14 shows the Pauling scale of electronegativity for the whole periodic table.

It can be estimated that the bond between two elements will be essentially ionic in character if the electronegativity difference between the element is greater than two. We could correctly predict an ionic bond between lithium and fluorine (electronegativity difference 3.0) and a largely covalent bond between carbon and chlorine (electronegativity difference 0.5).

FIG. 14. Values of electronegativity of the elements. (After N. N. Greenwood and A. Earnshaw, 1984)

The difference in electronegativity between hydrogen and fluorine is 1.8, so a fairly high percentage of ionic character is expected in the hydrogen fluoride molecule. One estimate gives a value of 45%. Some approximate values for electronegativity differences and percentage ionic character are: 1.0, 20%; 1.5, 40%; 2.0 60% and 2.5, 80%.

Polarizing power and polarizability

When a cation M^+ approaches an anion X^- it will tend to distort the electron cloud of the anion towards itself.

This *polarization* will lead to a build-up of electric charge between the nuclei and hence to some covalency in the bond. Now the importance of this covalency will depend on the *polarizing power* of the cation and the *polarizability* of the electron cloud of the anion. One aspect of polarizability is therefore an alternative to the use of electronegativity to predict the importance of covalency in a compound.

The polarizing power of a cation can be measured approximately by the ratio of its charge to its ionic radius, and is greatest for small, highly charged cations. Hence covalency will be more important in the series $B^{3+} > Be^{2+} > Li^+$ and $Li^+ > Na^+ > K^+$.

Similarly, large highly charged anions are more readily polarized than smaller anions, so we expect covalency to be more important in the series $P^{3-} > S^{2-} > Cl^-$ and $I^- > Br^- > Cl^- > F^-$. To be more specific, we expect greater covalency in compounds like boron sulphide, B_2S_3, than in potassium fluoride, KF.

Coordination number and oxidation state

As well as the valency, the terms coordination number and oxidation state are used to describe the state of an atom in a chemical compound.

The coordination number is simply defined as the number of nearest neighbours to the given atom in a chemical compound.

The oxidation state of an atom in a compound is defined as the charge (with the sign) which it would carry in the most probable ionic formulation of the compound. This is decided on the basis of the electronegativities of the elements in the compound. Take for example the covalent compound $SnCl_4$. The most probable ionic formulation would be $Sn^{4+}Cl_4^-$, so that the oxidation state for tin is IV and that for chlorine is $-I$. The compound may be written as tin(IV)chloride.

The coordination number of an atom often differs from its oxidation state. Thus in the ion BF_4^- the oxidation state for boron is III but the coordination number is four, while in the ionic caesium chloride (p. 22) the oxidation state for caesium is I but its coordination number is eight.

PROBLEMS

4.1 Arrange the following groups of elements in order of
   i) increasing first ionization energy,
   ii) increasing electronegativity.

     (a) B, N, F, Ne   (b) Mg, Ca, Sr, Ba, Hg

4.2 Arrange the following ions in order of decreasing ionic radius. Which ion will be formed least readily from its atom?

$$Li^+, Be^{2+}. B^{3+}, Na^+$$

4.3 Give the oxidation state and coordination number of the central atom in each of the following molecules or ions.

$$SiCl_4, SiCl_6^{2-}, AlCl_4^-, AlF_6^{3-}$$

5. Periodicity and the physical and structural properties of the main-group elements

An understanding of the physical properties of the elements (solid, liquid, or gaseous) requires some knowledge of their structures (if solid or liquid) and, more importantly, the magnitudes of the interatomic forces. The latter can be expressed in terms of the elemental binding energy, defined as the energy required to break down all the interatomic forces within the element, in its usual state, to form free gaseous atoms. So for iodine this would involve

$$I_2 (s) \rightarrow 2I(g).$$

Binding energies of the elements

The binding energies for the elements of the short periods are shown in Fig. 15. On crossing a period they increase with the number of valence (outer) electrons up to group IV and then fall to zero at the next noble gas. The noble gases are defined as having zero elemental binding energies simply because they exist as free atoms. At the other extreme carbon has the maximum number of bonding electrons. Thus in diamond each carbon atom is bound to four other carbon atoms in an infinite lattice; hence all four valence electrons on each carbon atom are engaged in bonding and a very high binding energy results. In comparison, boron has only three valence electrons. Nitrogen has five, but only three are engaged in forming the N≡N triple bond of the nitrogen molecule. As might be expected, boron and nitrogen have similar binding energies and each is lower than that for carbon. In the elements lithium and fluorine, each atom supplies only one bonding electron and the binding energies are low.

In general the elements with the greatest binding energies are unreactive while those with low binding energies are the most reactive. (The most notable exceptions to this rule are the noble gases which have zero binding energies but are most unreactive; this is readily explained in terms of the electron configurations of their atoms.) In order to gain further insight into the properties of other elements we must consider their structures in some detail.

Elements of groups I and II

The elements of groups I and II are all metals; they have the typical metallic lustre and ability to conduct heat and electricity. These properties

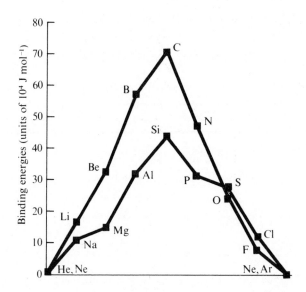

FIG. 15. Bindng energies of the elements of the short periods.

can be interpreted in terms of the structure and bonding in typical metals. As an example, we consider lithium metal to consist of a lattice of Li$^+$ ions. The 2s valence electrons of the lithium atoms are in a *valence band* which extends throughout the lattice, as illustrated in Fig. 16. The electrons are no longer associated with one lithium centre and are said to be *delocalized*.

The electrons can move freely throughout the lattice in the valence band and so bind each lithium ion to the other lithium ions in the lattice. This high electron mobility can also explain the high electrical and thermal conductivity shown by lithium and other metals.

For the group I elements the elemental binding energies are largely determined by the binding energies of the valence electrons in the individual atoms, that is by their ionization energies. This can be seen by comparing the plot of binding energies in Fig. 17 with that of ionization energies in Fig. 8.

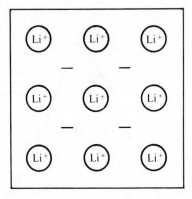

FIG. 16. Representation of the bonding in metallic lithium.

We might expect the melting points and boiling points of the metals to correlate with their binding energies, since both melting and boiling involve, to some degree, atomization of the elements. That this is indeed the case can be seen from Fig. 18 which contains a plot of binding energies and melting points for the group II elements. Similarly both binding energies and melting points of the group IIA elements are considerably higher than those of the group IA elements, because each group II atom provides two binding electrons to the valence band.

Reactivity of metals

When the metallic elements react they normally lose electrons and form cations; for example

$$2Li + F_2 \rightarrow 2Li^+F^-.$$

A similar process takes place when lithium reacts with water or with oxygen. Thus the reactivity of the elements in these reactions is likely to depend on the ionization energy of the elements atoms as well as on the elemental binding energy and the stability of the reaction products. This seems to be the case; thus the reactivity of the group IA elements increases on descending the group from $Li \rightarrow Cs$ as the ionization energy (and binding energy) decreases. Similarly the group IA elements are very

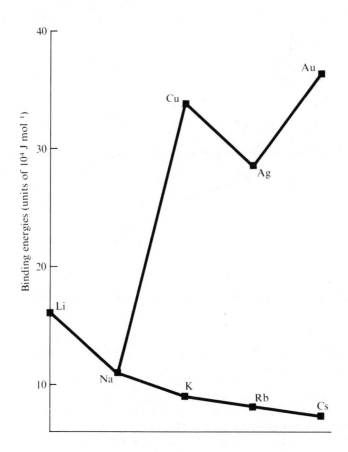

FIG. 17. Binding energies of the group IA and IB elements.

much more reactive than the group IB elements, which have considerably higher ionization energies and binding energies. Thus caesium reacts explosively with water or dilute acids whereas gold does not react at all.

Elements of groups III and IV

The binding energies for boron and carbon are very high and the elements are therefore hard and chemically unreactive. The bonding in the solids is

Fig. 18. Binding energies (●——) and melting points (▲---) of the group IIA and IIB elements.

somewhat different from the group IA and IIA elements, since the ionization energies required to form the ions B^{3+} and C^{4+} are very high, resulting in the binding electrons being more localized in covalent bonds. Boron and carbon, therefore, are not metallic. The structure of boron consists of a three-dimensional network of covalently bonded boron atoms, of which the basic building block is a B_{12} unit.

Boron is known as a *metalloid* because its properties are intermediate between those expected for a metal and for a non-metal. Thus it has a metallic lustre but is only a semiconductor of electricity. On descending group IIIB through the elements aluminium, gallium, indium, and thallium, the atoms increase in size and the binding energies and ionization energies decrease. The binding electrons then become more delocalized and have greater mobility so that the elements become increasingly metallic in character. Another consequence of the lower binding energies and ionization energies is that the heavier group IIIB elements are more reactive than boron towards water, oxygen, and the halogens, though they are considerably less reactive than the elements of groups IA and IIA.

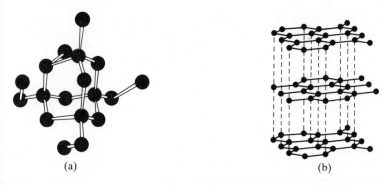

(a)   (b)

FIG. 19. Arrangement of atoms in the allotropes of carbon (a) diamond, (b) graphite.

A similar increase in metallic character is found on descending group IVB. Thus carbon has no metallic properties, silicon and germanium are metalloids having great importance as semiconductors of electricity, while tin and lead are more typical metals. The binding energies of all these elements are high, which partly explains the low chemical reactivity of the group IVB elements.

The elements, carbon and tin, can exist in more than one structural form, called allotropic forms. The allotropes of carbon, as shown in Fig. 19, are diamond and graphite, with graphite being the thermodynamically more stable allotrope at normal temperatures and pressures. Graphite is composed of planar, hexagonal layers, with each carbon atom covalently bonded to three others. The layers are stacked above one another, and weak π-bonding occurs between carbons of adjacent layers. The layers can slide because the interlayer forces are weak and this accounts for the lubricating properties of graphite.

In diamond, each carbon atom is tetrahedrally surrounded and covalently bonded to four equidistant neighbours, producing a three-dimensional network. This structure renders diamond the hardest known elemental substance. Carbon, in the form of diamond, is very unreactive at room temperature.

Graphite, although thermodynamically more stable then diamond, is more reactive, due to its more open structure. This is a good illustration of the relationship between structure and reactivity.

Silicon, germanium, and one allotrope of tin have structures like that of diamond, while the other allotropes of tin and lead have more typical metallic structures. Because of these different structures, the trends in melting points and boiling points for the group IV elements are less regular than those we saw for the elements of groups I and II.

Elements of groups V and VI

There is a steady increase in metallic properties on descending these groups, as for groups III and IV. However an important difference from earlier groups arises, as the lighter elements of groups V and VI exist as small, covalent molecules rather than as three-dimensional lattices. For instance nitrogen exists as N_2 molecules, each containing a nitrogen–nitrogen σ-bond and two nitrogen–nitrogen π-bonds. This triple bond is very strong so that nitrogen is an unreactive element, but although the binding energy is high the attractive forces between individual N_2 molecules (the intermolecular forces) are extremely weak. This has a profound effect on the physical properties of nitrogen which is a gas, and boils at -197°C at atmospheric pressure.

Because the heavier elements do not readily π-bond, they do not exist as small molecules but, instead, as larger molecular units. Phosphorus exists in several allotropic forms, the commonest allotrope is white phosphorus, which consists of symmetrical P_4 tetrahedral units in the solid, liquid, and gas phases (below 800°C).

Red and black phosphorus have highly polymeric structures in the solid state and one proposed structure for red phosphorus is shown below.

Arsenic, antimony and bismuth each exist in several allotropic forms, though the number of forms is not as extensive as for phosphorus. Arsenic is classified as a metalloid and in the gas phase exists as As_4 units. Bismuth exists only in a typical metallic lattice structure.

Similar trends are observed in group VI. Thus oxygen is a diatomic gas, O_2, containing double $O{=}O$ bonds (one σ-bond and one π-bond). Sulphur does not π-bond as readily as does oxygen and its most stable allotrope contains cyclic S_8 molecules which are crown-shaped.

However, other allotropic forms of sulphur may contain 6–20 atoms per ring.

Selenium and tellurium have chain structures and are semiconductors. The final element in group VI, polonium, is a more typical metal and is, for instance, a conductor of electricity.

Elements of group VII and group O

The group VIIB elements or halogens all consist of diatomic molecules with a single element–element bond, while the group 0 elements exist as the free atoms. The attractive forces between the halogen molecules or the noble gas atoms are largely determined by their *size*, and increase with increasing size of the atoms or molecules. In turn, the melting points and boiling points of the elements are determined by the magnitude of these interatomic or intermolecular forces (p. 34), and both of these physical properties increase in value on descending a group, as shown in Fig. 20.

None of the elements of these groups has any metallic properties. However while the group 0 elements are most unreactive, the group VIIB

FIG. 20. Boiling points (●——) and melting points (▲----) of the group VIIB and group 0 elements.

TABLE 1

The classification of the main-group elements

Metals		Metalloids			Non-metals		
						H	He
Li	Be	B	C	N	O	F	Ne
Na	Mg	Al	Si	P	S	Cl	Ar
K	Ca	Ga	Ge	As	Se	Br	Kr
Rb	Sr	In	Sn	Sb	Te	I	Xe
Cs	Ba	Tl	Pb	Bi	Po	At	Rn

elements are very reactive. Fluorine is the most reactive of all elements and will react with almost every other element in the periodic table. As the group I elements tend to lose an electron and attain a noble-gas electron configuration, the fluorine atom tends to gain an electron, forming the fluoride ion, in order to reach the noble-gas electron configuration of neon. We shall discuss the fluorides and other halides of the elements in Chapter 6.

Survey of the periodic table

The classification of the main-group elements as metals, metalloids, or non-metals is shown in Table 1. Bearing in mind that the transition elements and the lanthanide and actinide elements are all metals, it is clear that a high proportion of the chemical elements are metallic in nature and that this is especially pronounced for the heavier elements. Thus in the first short period of the periodic table there are two metals (Li and Be) and six other elements, while in the third long period there are thirty metals (Cs to Po) and only two non-metals (At and Rn).

PROBLEMS

5.1. Explain why boron and nitrogen have very different physical properties though their elemental binding energies are similar.
5.2. Arrange the following groups of elements in order of increasing and melting point.
    (a) Na, Mg, Al, Si, K   (b) P, S, Cl, Ar
    How do you explain these trends?
5.3. Arrange the following elements in order of their ability to conduct electricity.
    (a) Al, Si, P   (b) Se, Te, Po.

6. Periodicity and chemical compounds: the main-group elements

The chemical compounds of the transition metals and the lanthanides and actinides have certain characteristic properties because their ions have partially filled shells of d or f electrons. For this reason they are treated separately (in Chapters 7 and 8) from the main-group, or typical elements. Certain transition elements however do not show these characteristic properties. For instance zinc commonly forms a dipositive ion Zn^{2+} whose electron configuration is (Ar core) $3d^{10}$. The ion has a filled shell of d electrons and does not have characteristic transition-metal properties. For this reason the chemistry of zinc, and of some other elements which are formally transition elements, is included in this chapter.

In discussing the properties of chemical compounds we shall aim to treat classes of compounds which are formed by all elements of the periodic table. In this way periodic trends in the structure, bonding, and chemical properties of the compounds of the elements can be identified most readily.

We shall begin by discussing in some detail the properties of the halides of the elements, since these compounds illustrate very well the principles of chemical combination. Other classes of compounds can then be treated more briefly and with greater emphasis on trends in chemical behaviour.

Halides of the elements

Since the halogens are strongly electronegative elements (p. 28) their atoms normally carry some negative charge in their compounds with other elements; the magnitude of this charge depends on the electronegativity difference between the halogen and the particular element concerned. In *ionic halides* such as sodium chloride *all* of the bonding electrons are transferred to the halogen giving the halide ion.

$$2Na + Cl_2 \rightarrow 2Na^+Cl^-.$$

As we saw in Chapter 4, ionic compounds form lattice structures with strong interionic forces; they are therefore solids with high melting points and boiling points.

In *covalent halides* such as carbon tetrachloride the bonding electrons are localized in the carbon-chlorine bonds of individual CCl_4 molecules. The forces between these individual molecules—the intermolecular

TABLE 2

The fluorides of the elements in their group oxidation states

Gp IA	Gp IIA	Gp IIB	Gp IIIB	Gp IVB	Gp VB	Gp VIB	Gp VIIB
LiF	BeF_2		BF_3	CF_4			
NaF	MgF_2		AlF_3	SiF_4	PF_5	SF_6	
KF	CaF_2	ZnF_2	GaF_3	GeF_4	AsF_5	SeF_6	
RbF	SrF_2	CdF_2	InF_3	SnF_4	SbF_5	TeF_6	IF_7
CsF	BaF_2	HgF_2	TlF_3	PbF_4	BiF_5		
	Ionic			*Polymeric*		*Covalent*	

forces—are weak so that covalent halides are often gases, liquids, or low melting-point solids.

Thus the physical properties of element halides are largely determined by the nature of the bonding, and the classification into ionic or covalent halides is useful. A classification of the fluorides of the elements *in their group oxidation states* is shown in Table 2. Many of the elements form fluorides which do not readily fall into either ionic or covalent classification. These compounds often have polymeric structures in which molecules are associated through fluorine 'bridges'; they often have high melting points and boiling points but cannot accurately be described as ionic compounds, and are therefore classified separately as polymeric fluorides.

The usefulness of this classification is limited because the other halides are often very different from the fluorides, and because many of the elements form halides in oxidation states lower than the group number. However Table 2 does show some important trends. Thus it can be seen that on crossing the periodic table covalent compounds become more important, that the polymeric fluorides are generally formed by the heavier elements in a group, and that the first-row elements of groups V–VII do not form compounds in the group oxidation state. In order to explain these trends, and to compare the fluorides with the other halides of the elements, a more detailed survey of the properties of the element halides in groups is given in the following subsections.

Halides of the group 1A elements

The group IA elements readily lose their single valence electron to form monopositive cations (p. 17), and all of the halides of these elements are typical ionic compounds. As expected for ionic compounds they are all high melting-point solids; the melting points and boiling points of the sodium halides are shown in Fig. 21. It can be seen that they decrease

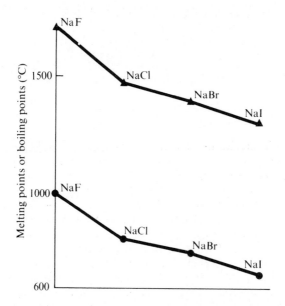

FIG. 21. Melting points (●──) and boiling points (▲──) of the sodium halides.

regularly as the size of the halide ion increases. We might have predicted this trend, for as the size of the halide ion increases the lattice energy decreases (p. 23). Since the lattice is partially or totally broken down when an ionic solid is melted or boiled, the melting point and boiling point also decrease, as less energy is required to break down the lattice.

These halides have been presented as purely ionic compounds, consisting of ions which are hard spheres. However, this is known to be an oversimplification, since even for LiF, NaCl, and KCl, experimental results have shown that there is some electron density transferred from the anion to the cation. As we have seen already, higher degrees of covalency are expected when either the cation or anion, in an 'ionic' compound, has a charge greater than one.

Halides of the group IIA and IIB elements

The group IIA metals have the electron configuration (noble gas core) ns^2. The extra energy needed to remove both valence electrons rather than just one is small compared with the extra lattice energy obtained by

TABLE 3

Melting points of the group II metal chlorides ($^{\circ}C$)

$MgCl_2$	$CaCl_2$	$SrCl_2$	$BaCl_2$
714	782	875	962
	$ZnCl_2$	$CdCl_2$	$HgCl_2$
	275	568	277

forming lattices with the dipositive cations such as Ba^{2+}, so that ionic halides MX_2 (for example $BaCl_2$ and $CaBr_2$) rather than MX are formed. Similarly the group IIB elements such as zinc, whose electron configuration is (Ar core) $3d^{10} 4s^2$, lose both ns electrons and form essentially ionic halides such as $ZnBr_2$.

The melting points of the group II metal chlorides are shown in Table 3. In group IIA there is a steady *increase* in melting point as the cation size increases, whereas a steady decrease is predicted on the basis of ionic lattice energies. The explanation of this effect is that the chlorides of the small cations such as Mg^{2+}, which have high polarizing power, are partially covalent and so have anomalously low melting points. The group IIB metal halides are also partially covalent and have low melting points. Thus mercury has the electron configuration (Xe core) $4f^{14} 5d^{10} 6s^2$, with the 6s electrons ineffectively shielded from the nuclear charge by the 4f and 5d electrons. Thus the ionization energy needed to form the Hg^{2+} ion is high and the mercury–halogen bonds in the mercuric halides have appreciable covalent character.

The beryllium(II) halides are typical covalent compounds having none of the properties expected of ionic salts. The first element of group II therefore differs considerably from the other elements; in this case the difference can be attributed to the small size of the beryllium atom. Either the ionization energy needed to form the Be^{2+} ion is too great, or the high polarizing power of the Be^{2+} ion leads to bonding which is largely covalent. We shall see that the first elements in subsequent groups often have some unique properties not possessed by the heavier members of the group.

In order for beryllium to form two covalent electron-pair bonds it is necessary to promote an electron from the filled 2s to a vacant 2p orbital

$$⬙ \quad ⬙ \quad ◯◯◯ \;\rightarrow\; ⬙ \quad ⬆ \quad ⬆◯◯ .$$

$$\text{1s} \quad \text{2s} \quad \text{2p} \qquad \text{1s} \quad \text{2s} \quad \text{2p}$$

In order to minimize repulsions between bonding electron pairs we would expect the beryllium halides such as $BeCl_2$ to be linear molecules.

$$Cl-Be-Cl$$

However there is a strong tendency for the beryllium atom to accept two more electron pairs into the remaining vacant 2p orbitals, thus attaining the neon electron configuration. This is accomplished in beryllium(II) chloride by the donation of a lone pair of electrons by each chlorine atom, giving in the solid state, a polymeric chain structure where

the notation $Cl \rightarrow Be$ represents an electron-pair bond in which both electrons are donated by the chlorine atom. The coordination number at each beryllium atom is four; this coordination number is very rarely exceeded for any element of the first short period. In the gas phase, $BeCl_2$ forms a bridged dimer

Halides of the groups III and IV elements
The trend towards increasing covalency in the element-halogen bonds on crossing the periodic table continues in these groups, and typical halides such as BF_3 and $SiCl_4$ are covalent molecules.

In order to form the trihalides of the group III elements an electron must be promoted.

The trihalides of aluminium, gallium, and indium dimerize, in the same way that beryllium chloride polymerizes, so that the metal atoms attain the noble-gas electron configuration. Thus aluminium(III) chloride, in

the gas phase, has the structure

$$
\begin{array}{c}
\text{Cl} \qquad \text{Cl} \qquad \text{Cl} \\
\diagdown \quad \vdots \quad \diagup \\
\text{Al} \qquad \text{Al} \\
\diagup \quad \blacktriangledown \quad \blacktriangledown \quad \diagdown \\
\text{Cl} \qquad \text{Cl} \qquad \text{Cl}
\end{array}
$$

In the crystalline state, aluminium(III) chloride has a lattice structure with 6-coordinate aluminium, which at its melting point, adopts the dimeric structure shown above. By contrast Al_2Br_6 and Al_2I_6 form dimeric units in the crystalline phase, as well as in the liquid and gaseous states.

The trifluorides of these elements are more strongly associated. Thus AlF_3 exists in an extended polymeric structure in which each aluminium atom is surrounded by six shared fluorine atoms. Hence aluminium attains a coordination number of six. The same coordination number is achieved when AlF_3, reacts with three fluoride ions to give the ion AlF_6^{3-}

$$AlF_3 + 3F^- \rightarrow AlF_6^{3-}.$$

We shall see that this coordination number of six is quite common, but is almost never found in the elements of the first short period, where the maximum coordination number, with very few exceptions, is four.

Exceptionally, the boron trihalides such as BF_3 do not dimerize or polymerize and have a trigonal planar structure

$$
\begin{array}{c}
\text{F} \\
| \\
\text{B} \\
\diagup \quad \diagdown \\
\text{F} \qquad \text{F}
\end{array}
$$

Boron trifluoride reacts with *one* fluoride ion to give the ion BF_4^- (the coordination number of boron is four—compare this with aluminium which forms AlF_6^{3-}).

In order for the group IV elements to form four covalent bonds it is again necessary to promote an s electron to a p orbital

$$ns \qquad np \qquad ns \qquad np$$

The elements achieve the noble-gas electron configuration by forming the four covalent bonds, so that there is less tendency for the halides to dimerize or polymerize. For carbon the maximum coordination number is reached in the carbon tetrahalides and these are quite unreactive

F<small>IG</small>. 22. Boiling points of some group IVB tetrahalides.

compounds. This is not true, however, for the tetrahalides of the other group IV elements, which have empty d orbitals capable of taking part in bonding. Thus silicon tetrachloride can accept electron pairs into its empty 3d orbitals and form ions such as $SiCl_5^-$ and $SiCl_6^{2-}$ in which the *coordination number* at silicone is five and six respectively.

This tendency to increase the coordination number above four is still greater for tin and lead, so that tin tetrafluoride (though not the other halides) polymerizes in order that the tin atoms may attain a coordination number of six.

Because of this polymeric structure, the melting point and boiling point for SnF_4 are considerably higher than for the other group IV tetrahalides, as shown in Fig. 22.

Lower oxidation-state halides. We have seen that in order for the group III and group IV elements to achieve their maximum oxidation state it is necessary to promote an electron from an s to a p orbital. The energy needed for this is usually more than offset by formation to two extra covalent bonds, but this may not always be the case. On descending a group the elements become larger and form weaker covalent bonds; gradually, the tendency to form the maximum oxidation state of three or four is decreased, and the tendency to form lower oxidation-state halides MX or MX_2 is increased. The latter tendency is greatest for the post-lanthanide elements thallium and lead for which the promotion energies are also high due to poor shielding of the 6s electrons. These elements form their most stable halides in an oxidation state two less than the group number, typical examples being TlBr and PbI_2. Formation of these lower oxidation-state compounds is referred to as the *inert pair* effect, since the ns^2 electron pair takes no part in bonding.

The higher oxidation-state compounds are most stable for the lightest halogens since these form the stronger covalent bonds. Thus TlF_3 is quite stable, but $TlCl_3$ decomposes readily to form TlCl and chlorine, while $TlBr_3$ and TlI_3 are still less stable. Similarly PbF_4 is stable, $PbCl_4$ is unstable, and $PbBr_4$ and PbI_4 cannot be prepared.

The lighter elements indium and tin also form isolable halides such as InCl and $SnBr_2$, but these are considerably less stable than the corresponding derivatives of thallium and lead and are readily oxidized to indium(III) or tin(IV) derivatives.

The ions Tl^+ and Pb^{2+} have much lower polarizing power than Tl^{3+} and Pb^{4+} so that the lower oxidation-state halides are the more ionic. Thus thallium(I) chloride, TlCl, is a typical ionic compound and is similar in many ways to the halides of the group IA elements. The bonding in $PbCl_2$ and in $SnCl_2$ is intermediate between covalent and ionic and these compounds resemble mercuric chloride in many of their properties.

Halides of the group V and VI elements

The electronegativities of the elements in these groups approach those of the halogens so that all the halides are covalent compounds.

Nitrogen and oxygen have the respective valence electron configurations

In each of these two cases, the 2s and three 2p orbitals 'mix' forming four sp^3-hybrid orbitals, as described previously for carbon (page 24). The hybridized nitrogen and oxygen atoms now bond to three and two fluorine atoms, respectively. NF_3 is pyramidal, and OF_2 has a bent structure, with the lone pairs occupying the other tetrahedral positions

The overall shapes of these molecules are greatly influenced by the lone pairs, and thus these lone pairs are said to be *stereochemically active*.

For both NF_3 and OF_2 the oxidation state is not the same as the group number but is eight minus the group number in each case. The heavier elements of these groups also form similar compounds, MX_3 in group V or MX_2 in group VI, but, since they have vacant d orbitals which may take part in bonding, they may also form higher oxidation-state halides.

Thus, by promoting an electron from the 3s to a vacant 3d orbital phosphorus may reach a valence electron configuration

having five unpaired electrons, so that five covalent bonds may be formed. The pentahalides PF_5, PCl_5, and PBr_5 are known, but PI_5 cannot be prepared. The stability of the pentavalent compounds decreases on descending the group until for bismuth only the rather unstable BiF_5 can be prepared. The pentahalides may add a further halide ion to give a six coordinate anion.

$$PF_5 + F^- \rightarrow PF_6^-$$

In group VI there are further possibilities. Thus for sulphur a single electron can be promoted from a 3p to a 3d orbital, allowing four covalent bonds to be formed.

If a 3s electron is also promoted to a 3d orbital a covalency of six is possible.

Examples of both types of compound are known and are more common than the simple two-covalent compounds. Thus sulphur forms SCl_4 and SF_6 as well as SCl_2, and similar compounds are formed by selenium and tellurium.

The formation of SF_6 is a further example of coordination number six for the elements of the second short period. On crossing the period, the isoelectronic series of ions AlF_6^{3-}, SiF_6^{2-}, PF_6^- is known, while in the first short period coordination number four is more typical as in the ions BeF_4^{2-}, BF_4^-, and CF_4.

Halides of group VII elements

As well as the simple halogens like I_2, several *interhalogen* compounds such as ClF, BrCl, or IBr can be formed. The heavier elements, perhaps by using their vacant d orbitals, can also form higher oxidation-state compounds, especially in combination with fluorine. Thus chlorine forms ClF_3, bromine forms BrF_3 and BrF_5, and iodine forms IF_5 and IF_7. It will be noticed that the highest oxidation states are observed for the *heaviest* elements, in contrast to the trends observed in groups III–V. In this group the stability of higher oxidation-state compounds appears to depend on the promotion energies, which steadily decrease on descending the group, rather than on the covalent bond strength.

Halides of group 0 elements

It was once thought that the noble gases would never form chemical compounds. However in 1962 the first such compound was isolated and several more have since been prepared. Thus xenon reacts directly with fluorine and XeF_2, XeF_4, and XeF_6 can be isolated. In order to form these compounds one or more electrons must be promoted from the 5p to the 5d orbitals of xenon. Like the ionization energy for xenon, this promotion energy is very high and stable compounds are only formed with the most powerful oxidizing agent fluorine. The promotion energies are even higher for the lighter noble gases so that argon forms no compounds with fluorine and krypton forms only KrF_2.

Reactions of halides with water

The group IA metal halides are mostly soluble in water. On dissolution the lattice energy of the crystalline halide is lost as the ions are separated, but extra energy called the solvation energy is gained by interaction of the ions with water. Since water, H_2O, is a polar molecule, it may interact with cations or anions as illustrated below for sodium and chloride ions.

$$Na^+ \cdots \overset{\delta-}{O} \overset{\diagup H^{\delta+}}{\underset{\diagdown H^{\delta+}}{}} \qquad Cl^- \cdots H^{\delta+} \underset{\diagdown \underset{\delta-}{O} - H^{\delta+}}{}$$

Sodium chloride is soluble in water partly because the energy gained by these electrostatic attractions is greater than the lattice energy which is lost.

The solvation energy for cations increases with their polarizing power, i.e. it increases with decrease in ionic size and with increase in charge. For the smaller group I cations (Li^+, Na^+) and all the group II metal cations the water molecules are tightly bound and may be retained when the halides are crystallized from aqueous solution. Thus beryllium bromide forms $BeBr_2 \cdot 4H_2O$ which contains the $Be(OH_2)_4^{2+}$ cation. Because of the stability of such hydrated salts, many anhydrous group II metal salts (for instance calcium chloride) are good drying agents.

Solutions of the halides of cations with still higher polarizing power are acidic because so much positive charge is developed on the hydrogen atoms of the attached water molecules that they may be lost as hydrogen ions. Thus aluminium(III) chloride forms an acidic solution in water due to the reaction

$$AlCl_3 \xrightarrow{\text{H}_2\text{O}} Al(OH_2)_6^{3+} \rightleftharpoons Al(OH_2)_5OH^{2+} + H^+.$$

The hydrogen ions are in fact better represented as H_3O^+ rather than H^+ since the free proton has a very high polarizing power (being very small) and interacts strongly with water. It is for this reason that the covalent hydrogen chloride ionizes in water

$$H_2O + HCl \rightarrow H_3O^+ + Cl^-.$$

Silicon tetrachloride gives a still more acidic solution than aluminium-(III) chloride, as it is completely hydrolysed by water to give silicic acid and hydrogen chloride[†]

$$SiCl_4 \xrightarrow{\text{H}_2\text{O}} SiO_2 \cdot xH_2O + HCl.$$

Phosphorous trichloride behaves like silicon tetrachloride, and elemental chlorine reacts with water to give a solution containing a mixture of hydrochloric and hypochlorous acids

$$Cl_2 \xrightarrow{\text{H}_2\text{O}} HCl + HOCl.$$

Compounds of hydrogen with the elements

The binary hydrides of the elements may be classified as shown in Table 4, according to the nature of their bonding: ionic, covalent or 'intermediate'. This classification is by no means perfect, since the precise nature of the bonding is not always known, and there is a gradation in properties between members of the various classes. For example, lithium hydride will certainly have some covalent character in the lithium–hydrogen bonds.

Ionic hydrides

The hydrides of the group IA and the heavier group IIA elements are ionic and contain the H^- or hydride ion. The hydride ion is rather reactive so that the ionic hydrides are reactive compounds. They react with water

[†] As noted on p. 46, CCl_4 is unreactive and is not affected by water.

TABLE 4

Classification of the element hydrides

Gp IA	Gp IIA	Gp IIB	Gp IIIB	Gp IVB	Gp VB	Gp VIB	Gp VIIB
LiH	BeH_2		BH_3	CH_4	NH_3	H_2O	HF
NaH	MgH_2		AlH_3	SiH_4	PH_3	H_2S	HCl
KH	CaH_2	ZnH_2	GaH_3	GeH_4	AsH_3	H_2Se	HBr
RbH	SrH_2	CdH_2	InH_3	SnH_4	SbH_3	HbH_3	HI
CsH	BaH_2			PbH_4	BiH_3		
Ionic		*Intermediate*			*Covalent*		

to give hydrogen according to the equation:

$$H^- + H_2O \rightarrow H_2 + OH^-.$$

The tendency towards covalency becomes more marked for the smaller elements, MgH_2, LiH, and particularly for BeH_2, which has a polymeric structure, with partially covalent beryllium–hydrogen bonds.

Covalent hydrides

The covalent compounds of hydrogen are usually gases or volatile liquids. The types of compound formed are often similar to the halogen derivatives but the following important differences may be noted.

(1) Hydrogen never causes an expansion of the valence electron shell beyond the stable octet. Thus with sulphur, hydrogen forms H_2S like SCl_2, but not H_4S or H_6S, though SCl_4 and SF_6 are easily prepared.

(2) Since the electronegativity of hydrogen is considerably lower than that of fluorine, hydrogen may develop a net positive or negative charge in its compounds while fluorine *always* develops a negative charge. Thus while in SiH_4 the bond polarity is $Si^{\delta+} -H^{\delta-}$, and in CH_4 there is almost no polarity in the C—H bonds, in H_2S or HCl the polarity is $H^{\delta+} -S^{\delta-}$ or $H^{\delta+} -Cl^{\delta-}$ This difference in polarity is reflected in the different chemical properties of these compounds. Thus SiH_4 is slowly hydrolysed by water to give hydrogen (like the ionic hydrides), while H_2S and HCl are acidic compounds since they react with water to give hydrogen ions:

$$HCl + H_2O \rightarrow H_3O^+ + Cl^-.$$

The hydrides of group IIIB, such as AlH_3, may add a further hydride ion to give an anion:

$$AlH_3 + Li^+H^- \rightarrow Li^+AlH_4^-.$$

Advantage may be taken of the high reactivity of the ionic hydrides in the preparation of covalent hydrides such as tin(IV) hydride by the reaction

$$2CaH_2 + SnCl_4 \rightarrow 2CaCl_2 + SnH_4.$$

Lithium aluminium hydride, $LiAlH_4$, reacts similarly with covalent halides and is often used to prepare covalent hydrides.

Discrete molecules of boron trihydride, BH_3, cannot be isolated under normal conditions, and the hydride exists as a gaseous dimer, B_2H_6.

Aluminium(III) hydride is an involatile, white solid with a polymeric structure.

Some discussion has already been given concerning the factors determining the shapes of molecules. The covalent hydrides of the elements in the first short period serve as a good illustration of these factors.

Molecular shape is governed by the need to minimize electrostatic repulsions between electron pairs, both bonding and non-bonding (lone pairs).

Monomeric BeH_2 possesses two bonding pairs of electrons and maximum separation of these pairs is achieved by a linear stereochemistry.[†] Monomeric BH_3, with three bonding pairs, adopts a trigonal planar shape,[†] whilst CH_4, with four bonding pairs, is tetrahedral. The NH_3 molecule possesses three bonding pairs of electrons and also a lone pair of electrons. According to the Valence Shell Electron Pair Repulsion theory (VSEPR), electrostatic repulsion between a lone pair and a bonding pair of electrons is greater than between two bonding pairs. This results in the HNH angle of the ammonia molecule being reduced from the tetrahedral angle 109.5 ° to 107 °. The NH_3 molecule is pyramidal. The water molecule, which possesses two lone pairs, adopts an angular or bent shape, with an HOH angle of 104 °.

The final hydride, HF, is of course linear.

[†] Remember that BeH_2 is normally polymeric, and BH_3 dimeric. The monomers exist only in the gas phase, at high temperatures.

The hydrogen bond

The melting and boiling points of the hydrogen compounds of the group IVB and group VIB elements are given in Fig. 23. It can be seen that, within a group, the boiling points increase on descending the group. The only exception is in group VIB where water has anomalously high melting and boiling points. The intermolecular forces between individual water molecules are greater than predicted, owing to the presence of *hydrogen bonding*. This effect is also present in ammonia, NH_3, and hydrogen fluoride, HF, which again have higher boiling points than might be expected. Thus, the hydrogen bond is only important when the hydrogen atom is bound to a small, electronegative element. It is thought to be essentially an electrostatic interaction and is usually represented by a

FIG. 23. Melting points (●——) and boiling points (▲---) of the compounds of hydrogen with the group IVB and VIB elements.

dotted line:

$$
\begin{array}{c}
\text{H} \\
\diagdown \\
\quad \text{O---H} \\
\diagup \qquad \diagdown \\
\text{H} \qquad\qquad \text{O—H.}
\end{array}
$$

Methyl derivatives of the elements

These compounds form part of an important class of compounds in which an organic group such as the methyl, CH_3-, or ethyl, $CH_3\cdot CH_2-$, group is bound to another element. If this element is a metal the compounds are known as organometallic compounds. They have many resemblances to the metal hydrides.

Thus the methyl derivatives of the elements of groups IA and IIA such as methyl-lithium CH_3Li and dimethylmagnesium $(CH_3)_2Mg$ are high melting-point polymeric solids with partially ionic metal-carbon bonds, while those of the group VB–VIIB elements are usually gases or volatile liquids with covalent element-carbon bonds.

In the derivatives of the elements of groups I–III there is a polarity in the metal–carbon bond in the sense $M^{\delta+}-CH_3^{\delta-}$, and they react with water to give methane and the metal hydroxide (compare this with the reaction of ionic hydrides with water, p. 51):

$$CH_3Li + H_2O \rightarrow CH_4 + LiOH;$$

$$(CH_3)_3B + 3H_2O \rightarrow 3CH_4 + B(OH)_3.$$

In the compounds of the group IVB elements, such as tetramethyl lead, $(CH_3)_4Pb$, there is little polarity in the metal–carbon bonds, while the carbon atoms develop some positive charge in the methyl compounds of elements in subsequent groups such as trimethylphosphine, $(CH_3)_3P$, dimethyl sulphide, $(CH_3)_2S$, or methyl iodide, CH_3I. None of these compounds is affected by water under normal conditions.

The boiling points of the methyl derivatives of the elements of groups IVB and VIB are shown in Fig. 24. They increase steadily on descending each group (compare this with the behaviour of the element hydrides in Fig. 23, p. 54).

Catenation

The vast majority of the elements exist as chains, layers, rings or three-dimensional networks of atoms. The formation of bonds between atoms of the same element is called catenation and this is the rule in the elemental state. Relatively few of the elements, however, form stable catenated *compounds*. Catenation is especially important for carbon, as

Fig. 24. The boiling points of the methyl derivatives of the group IVB and VIB elements.

the great variety of compounds containing chains of carbon atoms forms the basis for organic chemistry. Compounds with linear or cyclic chains of carbon atoms containing single, double, or triple carbon–carbon bonds may be formed as in ethane, CH_3-CH_3, cyclopropane, CH_2-CH_2 with CH_2 above, ethylene, $CH_2=CH_2$, or acetylene, $CH\equiv CH$. Very long chains can be formed as, for example, in polyethylene, $-(CH_2-CH_2-)_n$, where n can be $>10^3$.

On descending group IVB, catenated compounds become more difficult to make. The longest, well-characterized silicon hydride (silane) is Si_8H_{18}. A few cyclic silanes have been synthesised recently. The longest polygermane is Ge_9H_{20} and, for tin, no hydrides are known beyond Sn_2H_6. Lead forms only the unstable monomer, PbH_4. A similar trend is observed for the chlorides of group IVB. Carbon forms linear chlorides of almost indefinite length, whilst for silicon, the longest chain is in $Si_{10}Cl_{22}$. Germanium forms only Ge_2Cl_6 and tin and lead give no such compounds. Lead and tin exhibit weak catenation in methyl derivatives, such as $(CH_3)_6Pb_2$, having a lead–lead bond.

A similar trend is also apparent in group IIIB where only the lightest element, boron, forms an important series of catenated compounds which includes halides such as B_2F_4 and a great number of catenated hydrides. However in other groups this trend may be reversed. Thus in group VB catenated compounds of nitrogen, such as hydrazine, N_2H_4, are no more stable than those of phosphorus (for instance P_2F_4) or arsenic (for instance $As_2(CH_3)_4$), though the catenated compounds of antimony and bismuth are less stable; however in group VIB sulphur, with compounds such as H_2S_2, H_2S_5, and S_2F_{10}, definitely forms more stable catenated compounds than oxygen, whose simplest derivative is hydrogen peroxide, H_2O_2.

The elements of group VII show some tendency towards catenation forming polyhalide ions such as Cl_3^-, Br_3^-, I_5^-, I_7^-, and I_8^{2-}.

Overall the catenated compounds appear to be most common for elements of intermediate electronegativity. Thus carbon forms more compounds than the less electronegative elements of the remainder of group IVB and of groups IA, IIA, and IIIB. In group VIB oxygen is too electronegative and more stable compounds are formed by the less electronegative sulphur and selenium. A simple explanation of this effect is that the most electronegative and least electronegative elements prefer to form ionic or partially ionic bonds, so that they do not give stable catenated compounds which must contain completely non-polar covalent bonds.

Oxides of the elements

Oxygen is the most abundant element on earth and, like fluorine, it forms compounds with almost every other element. Since many of these have quite unusual properties, there is an extensive and varied chemistry of the compounds of oxygen. As a result, any classification tends to be either too simple to be reliable or too complicated to be useful. The classifications presented here are rather simplified but nonetheless are a good guide to the variety and periodicity in properties of oxides.

We shall discuss first the different ways in which oxygen can combine with the elements, and then examine periodic trends in structure and properties of the element oxides.

Ionic oxides

In forming an ionic oxide each oxygen atom gains two electrons, thus giving an O^{2-} ion. As a result of this high negative charge the lattice energies of ionic oxides are high, and ionic oxides are formed from all the elements which give ionic halides, despite the fact that formation of *isolated* O^{2-} ions from oxygen atoms is an energetically unfavourable process. Another effect of the high lattice energies is to cause the ionic oxides to be rather insoluble in water (since the solvation energy is often lower than the lattice energy), especially if the cation also carries a high charge. In fact the O^{2-} ion is unstable in aqueous solution so that the soluble ionic oxides react with water to form the corresponding hydroxides, according to the equation

$$O^{2-} + H_2O \rightarrow 2OH^-.$$

Covalent oxides

Covalent oxides are formed by non-metallic elements of groups IIIB-O. Carbon dioxide exists as a linear molecule, $O{=}C{=}O$, and each double bond comprises one σ-bond and one π-bond. The oxides of those elements which form similar double bonds with oxygen behave as expected for covalent, molecular compounds: they are usually gases, volatile liquids, or low melting-point solids (e.g., NO_2, SO_2). In contrast, if the element preferentially forms two single bonds with oxygen, polymeric oxides may result. For example, in silicon dioxide, SiO_2, each oxygen atom is bound to two silicon atoms, each of which in turn is bound to four oxygen atoms, thus giving the polymeric oxide quartz, whose structure is shown below.

● Silicon
○ Oxygen

TABLE 5

Classification of the element oxides according to structure

Gp IA	Gp IIA	Gp IIB	Gp IIIB	Gp IVB	Gp VB	Gp VIB	Gp VIIB	Gp 0
Li	Be		B	C	N	O	F	
Na	Mg		Al	Si	P	S	Cl	
K	Ca	Zn	Ga	Ge	As	Se	Br	
Rb	Sr	Cd	In	Sn	Sb	Te	I	Xe
Cs	Ba	Hg	Tl	Pb	Bi	Po	At	
		Ionic				*Polymeric covalent*		*Molecular covalent*

The different structures of CO_2 and SiO_2 result in quite different physical properties for these compounds. Thus CO_2 is a gas which sublimes at $-78\ °C$ at atmospheric pressure while SiO_2 is a solid which melts only at $2000\ °C$; this is another example of the first element in a group behaving quite differently from subsequent elements in the group. Because of these great differences in properties it is valuable to classify the covalent oxides as either small covalent oxides such as CO_2 or polymeric covalent oxides such as SiO_2. The full classification of the elements according to the nature of their oxides is shown in Table 5.

Oxides of the group I–IV elements

The oxides of these elements are given in Table 6. With the exception of carbon, all the elements form ionic or polymeric covalent oxides.

The solubility of the oxides in water is strongly dependent on the size and charge of the cation in the ionic oxides, and is greatest for large cations carrying a low charge. Thus on crossing a period the solubilities decrease, so that Cs_2O and BaO are easily soluble in water (with reaction, p. 58) while HgO, Tl_2O_3, and PbO_2 are insoluble. Similarly, on descending a group the solubilities increase, so that MgO is insoluble in water while BaO dissolves readily.

The heavier elements of groups IIIB and IVB form oxides both in the group oxidation state (for instance $Sn(IV)O_2$) and in the oxidation state two less than the group number (for instance $Sn(II)O$). The relative stability of the lower and higher oxidation state oxides is similar to that of the corresponding element fluorides (p. 47). Carbon, the first element in group IV, differs completely from the other elements in these groups in forming the molecular covalent oxides carbon dioxide, CO_2, and carbon

TABLE 6

The simple oxides of the elements of groups I–IV

Gp IA	Gp IIA	Gp IIB	Gp IIIB		Gp IVB	
Li_2O	BeO		B_2O_3		CO_2	CO
Na_2O	MgO		Al_2O_3		SiO_2	
K_2O	CaO	ZnO	Ga_2O_3		GeO_2	
Rb_2O	SrO	CdO	In_2O_3	In_2O	SnO_2	SnO
Cs_2O	BaO	HgO	Tl_2O_3	Tl_2O	PbO_2	PbO

monoxide, CO. As we have seen this is due to the ability of carbon to form strong π-bonds with oxygen.

Oxides of the group V elements

All of the elements of this group form oxides of the type $E_2O_3(E = $ group V element) and all but bismuth also form the oxides E_2O_5. However the structures and properties of these oxides are often quite different.

Nitrogen also forms a unique series of oxides which includes dinitrogen monoxide, N_2O, nitrogen oxide, NO, nitrogen dioxide, NO_2, and dinitrogen tetroxide, N_2O_4, all of which are small covalent molecules in the gas phase. Some of these oxides contain an 'odd' or unpaired electron. Thus we may write the electronic structure of nitrogen dioxide as

$$\overset{..}{N}\!:\!:\!\overset{..}{O}: \qquad \overset{.}{N}\!=\!O$$
$$:\!\overset{..}{O}: \qquad or \qquad \overset{\downarrow}{O}$$

with an odd electron on nitrogen. The nitrogen atom can gain the stable octet of electrons if nitrogen dioxide dimerizes, and indeed nitrogen dioxide exists in equilibrium with its dimer, dinitrogen tetroxide:

$$O_2N\cdot \; + \; \cdot NO_2 \rightleftharpoons O_2N - NO_2.$$

The oxides of phosphorus, which are formed by burning the element in air, are best written as P_4O_6 and P_4O_{10}, since the solid compounds contain the molecular units:

It can be seen that the tetrahedral P_4 units present in yellow phosphorus are maintained, with oxygen atoms bridging between pairs of phosphorus atoms. In P_4O_{10} each phosphorus is bound to an additional 'terminal' oxygen atom.

Thus the oxides of phosphorus are on the borderline between the polymeric covalent oxides, with infinite networks of atoms, and the small, covalent oxides. The oxides of arsenic and antimony are similar, but the only stable oxide of bismuth, Bi_2O_3, is an ionic compound.

The nitrogen analogues of these phosphorus oxides are N_2O_3 and N_2O_5. N_2O_3 exists as discrete, covalent molecules, at low temperature, whilst N_2O_5, in the solid state, consists of an ionic array of $[NO_2]^+$ and $[NO_3]^-$ ions.

Oxides of the group VIB, VIIB, and 0 elements

The well-characterized oxides of these elements are listed in Table 7.

The group VIB elements form an important group of oxides. Oxygen, like fluorine, always causes an expansion of the valency shell of the heavier group VIB elements beyond the octet, so that sulphur forms the oxides sulphur dioxide, SO_2, and sulphur trioxide, SO_3, but *not* sulphur monoxide, $S{=}O$, which would be a sulphur analogue of the oxygen molecule. On descending the group the following trends may be seen:

(1) The stability of element–oxygen double bonds decreases so that sulphur dioxide and sulphur trioxide contain $S{=}O$ double bonds and are volatile (the vapours contain discrete SO_2 or SO_3 molecules), while SeO_2 and TeO_2 are solid polymeric oxides. Furthermore, the coordination number of the group VIB element increases down the group: $SO_2,2$; $SeO_2,3$; $TeO_2,4$; $PoO_2,8$.

(2) The stability of the oxidation state VI decreases, so that SeO_3 and TeO_3 are much less stable than SO_3.

The group VIIB elements form a wide range of oxides, as shown in Table 7, but all are unstable compounds and many are explosive. No

TABLE 7

Oxides of the elements of groups VIB, VIIB, and 0

Gp VIB		Gp VIIB						Gp 0	
O_2		F_2O							
SO_2	SO_3	Cl_2O	Cl_2O_3	ClO_2	Cl_2O_4	Cl_2O_6	Cl_2O_7		
SeO_2	SeO_3	Br_2O		BrO_2					
TeO_2	TeO_3			I_2O_4	I_4O_9	I_2O_5		XeO_3	XeO_4
PoO_2									

meaningful periodic trends can be found. Xenon forms two binary oxides, XeO_3 and XeO_4. The former oxide is an explosive solid, formed by hydrolysis of XeF_6. XeO_4 is an explosively unstable gas, prepared by the addition of Ba_2XeO_6 to concentrated sulphuric acid.

Acidic and basic oxides

The classification of the oxides of the elements as acidic or basic is dependent on their reactions with water. It is well-known that water undergoes a self-ionization process:

$$2H_2O \rightarrow H_3O^+ + OH^-.$$

In water the product $[H_3O^+][OH^-]$ is a constant[†] and in pure water $[H_3O^+] = [OH^-] = 10^{-7}$M. at 25 °C. If addition of another substance to water causes an increase in $[H_3O^+]$ that substance is an acid while if it causes an increase in $[OH^-]$ it is a base.

Thus sodium oxide is a basic oxide since it dissolves in water according to the equation

$$Na_2O + H_2O \rightarrow 2Na^+ + 2OH^-,$$

and causes an increase in hydroxide ion concentration.

On the other hand, when sulphur trioxide, SO_3, reacts with water there is no tendency for a similar ionization to occur giving the S^{6+} ion. Rather, the sulphur atom, having a high polarizing power, accepts a further electron pair from an oxygen atom of a water molecule into a vacant d orbital and gives sulphuric acid according to the equation

$$SO_3 + 3H_2O \rightarrow SO_4^{2-} + 2H_3O^+.$$

Thus the oxides Na_2O and SO_3 show extreme differences in their reactions with water, the former giving a strongly basic solution and the latter a strongly acidic solution. Basic oxides and acidic oxides react with one another to give salts, for example

$$Na_2O + SO_3 \rightarrow Na_2SO_4.$$

Certain oxides react with both basic and acidic oxides in solution. For instance, although aluminium(III) oxide is itself too insoluble to react readily with either acids or bases, aluminium(III) hydroxide, $Al(OH)_3$, dissolves both in sulphuric acid to give aluminium(III) sulphate

---

[†]The brackets represent concentrations. Thus $[H_3O^+]$ is a shorthand way of writing concentration of H_3O^+ ions.'

Table 8

Classification of the oxides

Gp IA	Gp IIA	Gp IIB	Gp IIIB	gp IVB	Gp VB	Gp VIB	Gp VIIB
Li	Be		B	C	N	O	F
Na	Mg		Al	Si	P	S	Cl
K	Ca	Zn	Ga	Ge	As	Se	Br
Rb	Sr	Cd	In	Sn	Sb	Te	I
Cs	Ba	Hg	Tl	Pb	Bi	Po	At

Basic oxides   Amphoteric   Acidic oxides
                                   oxides

(remember that the ions H^+ and Al^{3+} are in fact strongly hydrated, p. 51),

$$Al(OH)_3 + 3H^+ \rightarrow Al^{3+} + 3H_2O,$$

and in sodium hydroxide to give sodium aluminate,

$$Al(OH)_3 + 3OH^- \rightarrow Al(OH)_6^{3-}.$$

The former is a reaction typical of basic oxides or hydroxides while the latter is typical of acidic oxides. Oxides like aluminium oxide, which show both acidic and basic properties, are called *amphoteric* oxides.

We may therefore further classify the elements according to whether their oxides are basic, acidic, or amphoteric. This classification is given in Table 8, and the similarities with the classification in terms of structure and bonding in Table 5 (p. 59) are obvious. Thus the ionic oxides are usually basic though, if the central metal carries a high charge as in group III, they may be amphoteric. The polymeric covalent oxides show either amphoteric or acidic properties, while the molecular covalent oxides are almost always acidic.

The basicity of the main group metal oxides increases on descending a group of the periodic table (e.g., $BeO < MgO < CaO < SrO < BaO$), though the reverse trend may be observed for the later transition metal oxides. The oxidation state of the metal also has a bearing on the acid/base character of the oxide. The acidity increases with increasing oxidation state (e.g., the acidity of $MnO < Mn_2O_3 < MnO_2 < Mn_2O_7$).

Like the parent acidic oxides, the oxyanions such as the sulphate ion, SO_4^{2-} (from SO_3), or the carbonate ion, CO_3^{2-} (from CO_2), may exist as molecular ions with multiple element–oxygen bonds or as polymeric ions with bridging oxygen atoms. Thus the molecular covalent oxide CO_2 gives the molecular carbonate ion CO_3^{2-}, while the polymeric oxide SiO_2 always gives polymeric silicates, which, like silicon dioxide itself, are highly insoluble and are often present in rocks.

Sulphides of the elements

The element sulphides can be classified, like the oxides, as ionic, molecular covalent, or polymeric covalent sulphides. However, for a number of reasons, the periodic distribution of these structural types is quite different from that for the oxides.

Ionic sulphides

The only truly ionic sulphides are those of the group IA and group IIA elements. Thus there are far fewer ionic sulphides than there are ionic oxides. Several factors may account for this difference.

(1) Since the S^{2-} ion is larger than the O^{2-} ion, lattice energies for ionic sulphides will be lower than for oxides, so that they are less likely to be formed.

(2) The larger S^{2-} ion is much more polarizable than the O^{2-} ion so that an increased tendency to form covalent compounds is expected. This trend is also anticipated from the fact that sulphur is less electronegative than oxygen.

One feature of the ionic sulphides, not seen for the oxides, is the tendency to form polysulphide ions, S_n^{2-}, by catenation. Thus, for potassium there exist four characterized polysulphides K_2S_n ($n = 3$–6). The polysulphide anions have an open chain structure (e.g., $^-S \diagup^{\textstyle S} \diagdown S^-$).

Covalent sulphides

Sulphur forms stable double bonds only to the elements oxygen and carbon. Thus the only simple molecular sulphide is formed by carbon which gives carbon disulphide, CS_2, a volatile liquid which has the structure $S{=}C{=}S$.

Since there are fewer ionic sulphides and molecular covalent sulphides in comparison with the oxides, it follows that there are more polymeric covalent sulphides. However these polymeric sulphides exist in a wide variety of structures, so that it is difficult to find meaningful periodic trends and we shall not discuss them further.

The diagonal relationship

We have already seen the value of the concept of the polarizing power of a cation in interpreting the chemical and physical properties of chemical compounds. Thus we have discussed covalency in metal halides, hydrides, oxides, and sulphides and the reactions of metal oxides and halides with water in terms of the polarizing power of the cation. In general, we expect compounds of cations with similar polarizing power to have similar properties.

Now we have seen (p. 30) that the polarizing power of a cation can be measured approximately by the ratio of its charge to its radius, and it can readily be calculated that elements on diagonal lines in the periodic table, as shown below, will give cations having similar polarizing powers.

Thus the ion Al^{3+} is larger but carries a higher charge than the ion Be^{2+}, so that these ions have similar polarizing powers, and we expect similarities in the chemistry of these elements. This is indeed the case and, in fact, beryllium resembles aluminium more closely than it resembles magnesium or the other group II elements. Both beryllium and aluminium are metallic elements; they are not attacked by air or water, owing to the formation of an inert oxide film on the metallic surface; both react with strong bases to give respectively beryllates and aluminates. Both elements form hard, unreactive, very high melting-point oxides and amphoteric hydroxides which are insoluble in water. The anhydrous halides of both elements are covalent with bridging halogen atoms; solutions of the halides in water are acidic and the hydrated salts are obtained on crystallization. Be_2C and Al_4C_3 are believed to be the only common salt-like metal carbides that yield methane on hydrolysis.

The chief difference between the elements is that aluminium can form compounds with coordination number six while the maximum coordination number for beryllium, as for other elements of the first short period, is four. Thus in AlF_3 each Al atom is six coordinate while in BeF_2 the Be atoms are four coordinate. On crystallization from water aluminium salts usually contain the hydrated cation $Al(OH_2)_6^{3+}$ while beryllium gives only the $Be(OH_2)_4^{2+}$ ion.

Lithium resembles magnesium less strongly, and the differences can be attributed to the greater polarizing power of Mg^{2+} compared with Li^+. Thus lithium salts are more typically ionic than magnesium salts; in aqueous solution the Li^+ ion is solvated less strongly than Mg^{2+}, and lithium hydroxide is a stronger base than magnesium hydroxide.

There are strong similarities between the chemistries of boron and silicon. Thus both elements are metalloids; they give covalent monomeric chlorides which are completely hydrolysed by water to give boric acid or silicic acid and hydrogen chloride, and both form acidic oxides.

More generally, we have seen that the metalloid elements occur on a diagonal line across the periodic table (Table 1, p. 40) as do the polymeric covalent oxides (Table 5, p. 59) or amphoteric oxides (Table 8, p. 63), as well as the hydrides with intermediate properties between the typical ionic and covalent hydrides (Table 4, p. 52).

This diagonal relationship is most useful for the elements of groups I–IV, and is much less important for the most electronegative elements, which have little or no tendency to form cations. Thus there is little resemblance between the compounds of carbon and phosphorus or between those of nitrogen and sulphur.

Anomalous Behaviour of the First Row Elements

The discussion above emphasised *similarities* in properties between a first row element and a second row element from the next group (such as the pair of elements lithium and magnesium). In this section, a complementary discussion of the *differences* between the first row element and its heavier congeners will be given, and should serve as a summary and review of the properties of main group elements and their compounds.

Reactivity of first row metals and metalloids

We expect that the reactivity of a metal towards water will depend, in most cases, on its ionization energy and binding energy. This is because reaction with water involves oxidation of the metal. The first row elements have higher ionization energies and binding energies than their heavier congeners and so are less reactive. The group IA elements react with water according to the equation (M = Li, Na, K, Rb or Cs):

$$2M + 2H_2O \rightarrow 2M^+ + 2OH^- + H_2.$$

Lithium reacts with water only slowly at 25 °C whereas the other members of group IA react extremely vigorously. Similarly, beryllium is much less reactive than magnesium and the heavier group IIA metals, though this is partly due to formation of an inert film of BeO on the surface, which prevents further reaction. Beryllium is not even attacked by steam at red heat. Crystalline boron is much less reactive than aluminium and, for example, is not attacked by boiling hydrochloric acid. Many other reactions of the metallic elements involve their oxidation and similar reactivity trends are observed.

Metallic character of first row elements

We have seen that the first row elements are less metallic than their heavier congeners. The first row elements of groups Ia and IIA, Li and

Be, are still metals and differences in metallic properties from the heavier elements are not immediately obvious. They are less highly conducting due to the higher ionization energies, but are still clearly metallic in nature. However, at boron the energy needed to ionize boron to B^{3+}, with the electrons going into a metallic conduction band, is too high and boron instead adopts a covalent polymeric structure with more localized boron–boron bonds. Boron then has the properties of a metalloid. At carbon, this trend continues and carbon is a non-metal and a very poor conductor of electricity. The heavier elements of groups IIIA and IVA become increasingly metallic as the group is descended, due to a decrease in the ionization energies and hence a more delocalized form of bonding. Aluminium, for example, is a true metal. The strong covalent bonds formed in the elemental forms of boron and carbon are also present in many of the compounds of these elements, for both of which catenation is a very important feature.

Covalent character for first row element compounds

The ions Li^+, Be^{2+} and B^{3+} are very small and hence have higher polarizing power compared to heavier element ions (e.g. Na^+. Mg^{2+}, Al^{3+}). As a result, the compounds of these first row metallic or metalloid elements are more covalent than the corresponding compounds of the heavier congeners. We have already seen examples of this effect, for instance in the polymeric covalent chain structure of $BeCl_2$ compared to the ionic lattice structures for $MgCl_2$, $CaCl_2$, $SrCl_2$ and $BaCl_2$. For the group IA elements, the effect is seen particularly clearly for the organometallic compounds. Thus, methyllithium, CH_3Li, is more covalent, and hence more stable and much less reactive than the more ionic CH_3Na or CH_3K. Similarly, in group IIIA the hydride of boron, B_2H_6, is a covalently bonded dimer whereas AlH_3 is a polymer with bridging hydrogen atoms, and the Al–H bonds are certainly more polar in the sense $M^{\delta+}$ —$H^{\delta-}$ than for boron. In group IVA a very similar trend is seen with carbon forming less polar covalent bonds than silicon, germanium, tin and lead.

Lower coordination numbers for first row elements

The first row elements, with very few exceptions, have a maximum coordination number of four, whereas the second and third row elements often expand their coordination sphere to bond to five or six other atoms. This is most simply explained by the availability of empty 3d orbitals for the heavier elements, although theoretical chemists suggest that this explanation is not the whole truth. Many examples have already been given and they will not be repeated here.

Multiple bonds for first row elements

The ability of the first row elements to form strong π-bonds as well as σ-bonds, has a remarkable impact on the properties of the elements nitrogen and oxygen as well as on the compounds of carbon, nitrogen and oxygen. We have seen that nitrogen, N≡N, and oxygen, O=O, are gases under normal conditions whereas the heavier congeners, which form σ-bonded oligomers or polymers with σ-bonding only, are solids at room temperature. In chemical compounds, we have seen that carbon dioxide, O=C=O, with multiple C=O bonds, is normally a gas while SiO_2 and the heavier group IVA element oxides are high-melting solids. Ethylene, $H_2C=CH_2$, is a gas, which is used in vast quantities in the petrochemical industry, but $H_2Si=SiH_2$ is known only as a very shortlived intermediate in the gas phase. The C—C π-bond energy is estimated to be 198 kJ mol$^-$, while the Si—Si π-bond energy is only about 90 kJ mol^{-1}, due to poor sideways overlap of the larger and more diffuse $3p\pi$ orbitals of silicon, thus accounting for the instability of the silicon compound.

Oxidizing ability of oxygen and fluorine

These elements are much stronger oxidizing agents than the heavier elements of groups VIB and VIIB. This is due to the high electron affinities and electronegativities of the elements, to the strong bonds formed to other elements and, for fluorine, due to the weak F—F bond in elemental fluorine. Thus oxygen and, especially, fluorine react with most of the other elements in the periodic table and both are good at stabilizing high oxidation states of both main group and transition elements. Some examples with fluorine include formation of IF_7, BiF_5, AgF_2 and AuF_5 and oxygen and fluorine are the elements which have an extensive chemistry with the noble gases, forming compounds such as XeO_3 and XeF_6.

PROBLEMS

6.1. Explain why in covalent compounds bromides usually have higher boiling points than chlorides, while in ionic compounds the reverse is true.

6.2. Which of the following halides do you expect to be hydrolysed rapidly by water?

$$BCl_3, \ CCl_4, \ SiCl_4, \ PCl_5, \ SF_6$$

6.3. An element is a metalloid. What properties do you expect its halides, hydride, and oxide to possess? (Hint: compare Tables 1, 2, 4, 5, and 8.)

7. Periodicity and chemical compounds: the transition elements

Introduction

Many compounds of the transition elements have certain properties which are not possessed by compounds of the main-group elements, owing to the presence of partially-filled shells of d electrons on the transition-elements ions.

However there are also similarities between the transition elements and the main-group elements, and, before treating the compounds of the transition elements in more detail, it will be valuable to make a brief general survey of the similarities and differences between the chemistries of the elements. Since the transition elements are all metallic in nature, the most useful comparisons can be made with the metallic main-group elements of groups I–IV.

The consecutive ionization energies of the elements of the first long period from potassium to germanium are shown in Fig. 25. There is usually an obvious 'break' in the separations between consecutive ionization energies for a main group element. Thus for potassium there is an obvious break between the first and second ionization energies, so that K^+ is easily formed but K^{2+} is never present in chemical compounds. Similarly the calcium atom easily loses two electrons to reach the (argon) noble-gas electron configuration, but a prohibitive amount of energy would be required to form the Ca^{3+} ion. For the group IIIB element gallium this break, which is less marked than for the group IA or IIA elements, comes between the third and fourth ionization energy so that gallium prefers to adopt the group oxidation state of three.

These trends are still present in the early part of the transition series. For the group IIIA element scandium there is an obvious break between the third and fourth ionization energies and, as we might expect, the chemistry of scandium is dominated by the tendency to form the Sc^{3+} ion. This ion has *no* d electrons so that compounds of scandium(III) show few typical 'transition' properties.

On passing across the transition series the separations between subsequent ionization energies increase and it becomes progressively more difficult to form transition-metal ions in their group oxidation states. Therefore compounds of titanium(IV) are more stable than those of vanadium(V), which in turn are more stable than those of chromium(VI) or manganese(VII), while no compounds of iron(VIII) are known. It

Fɪɢ. 25. Ionization energies of the elements of the first long period. Consecutive ionization energies increase so that the first ionization energies lie on the bottom line and the fourth ionization energies (where known) on the top line.

follows that the formation of stable compounds in oxidation states lower than the group number becomes increasingly important after scandium.

Thus titanium forms not only the chloride $TiCl_4$ but also $TiCl_3$ and $TiCl_2$. In $TiCl_4$ the electron configuration at titanium(IV) is (Ar core) and there are no d electrons. Titanium tetrachloride has many similarities with the group IVB tetrachlorides such as $SnCl_4$; both are colourless liquids which are violently hydrolysed by water. However, in $TiCl_3$ or $TiCl_2$ the electron configuration at titanium is respectively (Ar core) $3d^1$ and (Ar core) $3d^2$. It is the presence of these odd d electrons which leads to the characteristic colours and magnetic properties of the transition-element compounds. Thus $TiCl_3$ is a violet solid and $TiCl_2$ is a black solid; both are quite different from $SnCl_2$, which is white.

The typical 'transition' property of variable oxidation state then arises for two reasons: first because it is often difficult to reach the group oxidation state, and secondly because the differences in stability between the various lower oxidation-state compounds are small. The stable oxidation states often vary in units of one, rather than in units of two as found for the main-group elements. Thus, as noted above, titanium often forms compounds in oxidation states IV, III, and II, while tin forms compounds in oxidation states IV and II only.

On crossing the transition series each successive element can form compounds in a wider range of oxidation states. Thus manganese compounds in all oxidation states from −I to VII are known, but after manganese the elements iron, cobalt, and nickel never form compounds in the group oxidation state, so that for these elements some of the d electrons never take part in bonding and are effectively core electrons. In fact on crossing the transition series it seems that the 3d orbitals are stabilized with respect to the 4s and 4p orbitals, as shown diagrammatically in Fig. 26, and by the end of the transition series they have entered the electron core.

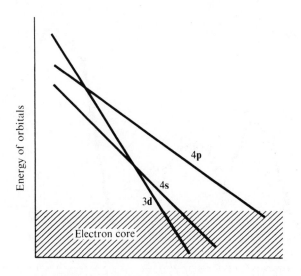

FIG. 26. The stabilization of the 3d orbitals on crossing the transition series.

The elements of group IIB at the end of the transition series very rarely use there d orbitals in bonding and, as we have seen (p. 41), they show no transition-metal properties. Overall then, the first elements and the last elements of the transition series behave like main-group metallic elements, but the inner transition elements, which commonly form ions having partially-filled shells of d electrons, have the extra property of forming compounds, which are often highly coloured, in a wide range of oxidation states.

Binding energies of the transition elements

The elemental binding energies (p. 32) of the transition elements again show evidence that the 3d electrons enter the electron core towards the end of the transition series. Binding energies for the first and third transition series are given in Fig. 27. It can be seen that at the beginning of the transition series the binding energies increase with the number of electrons, indicating that all the d electrons contribute to the binding. However towards the end of each series there is a drop in binding energy as progressively more d electrons enter the electron core and no longer contribute to the binding. At the end of the transition series (Group IIB) only the 4s electrons contribute to the binding energy, which is so low at

FIG. 27. Binding energies for the elements of the first and third transition series.

the end of the third transition series that mercury is a liquid at room temperature. This may be described as an example of the inert-pair effect (p. 48). It is because the 3d electrons have entered the electron core that the group IIB elements show no typical transition-metal properties.

We can gain our first insight into another general trend in transition-metal chemistry by comparison of the binding energies for the first and third transition series. In the first transition series the general increase in binding energies ends at vanadium, and vanadium is the last element to form stable compounds in the group oxidation state. Vanadium(V) compounds are quite stable, while subsequent elements in their group oxidation states, for instance Cr(VI), are strongly oxidizing, as they are easily reduced to lower oxidation states. In the third transition series the general increase in binding energies on crossing the periodic table ends at group VIA, and it seems that for tungsten all the 5d electrons contribute to the binding. All the 5d electrons can also be used in forming compounds with other elements, so that tungsten(VI) compounds are quite stable compared with those of chromium(VI).

This increased stability of high oxidation states for the elements of the second and third transition series compared with the first is an important feature of transition-metal chemistry.

FIG. 28. Melting points of the elements of the first and third transition series.

Finally in this section, we note that the binding energies can be correlated with many of the physical properties of the transition elements. The melting points of the first and third transition series, shown in Fig. 28, correlate very well with the binding energies in Fig. 27. We saw similar correlations for the main group elements.

Oxidation states of the transition elements

We have already predicted some of the trends in oxidation states of the transition elements. Let us now examine these trends more closely. In Table 9 are listed those chlorides of the transition metals which are sufficiently stable to be isolated. The property of multiple oxidation states of the transition metals is well illustrated by the great number of compounds formed. We also see that higher oxidation states are favoured by the second and third transition-series elements.

Another feature is the evident high stability of the oxidation state II for the chlorides of the first transition series. We might have predicted this from the ionization energies in Fig. 25 (p. 70). The third ionization energies for the first transition-series elements are much higher than the first two ionization energies. To form the dipositive ions the two 4s electrons are generally lost and the ionization process is

$$[\text{Ar core}]\ 3d^n4s^2 \rightarrow [\text{Ar core}]\ 3d^n.$$

The 4s electrons are removed *before* the 3d because the latter are strongly stabilized as the effective nuclear charge increases. This stabilization also means that a high energy is required to remove a further electron to form the tripositive ion.

In the second and third transition series the stabilities of the 4d and 5d orbitals are not so dependent on the nuclear charge, so that third and subsequent ionization potentials are lower than for the first transition series; this can be seen by comparing the ionization energies for the second transition series in Fig. 29 with those for the first transition series in Fig. 25 (p. 70). For this reason the oxidation state II is not so common for the heavier transition elements.

High oxidation states

As is the case with the main group elements, the highest oxidation states are attained when the transition metals combine with fluorine or oxygen. The highest oxidation-state fluorides of the transition elements are shown in Table 10. Group oxidation states are commonly found up to group VIIA in which rhenium forms of ReF_7. Thereafter high oxidation states become progressively less stable as the d orbitals enter the electron core.

In combination with oxygen yet higher oxidation states may be attained. Thus in the first transition series the group oxidation state is achieved by

TABLE 9

Chlorides of the transition elements

Gp IIIA	Gp IVA	Gp VA	Gp VIA	Gp VIIA	Gp VIII	Gp VIII	Gp VIII	Gp IB	Gp IIB
$ScCl_3$	$TiCl_2$ $TiCl_3$ $TiCl_4$	VCl_2 VCl_3 VCl_4	$CrCl_2$ $CrCl_3$	$MnCl_2$	$FeCl_2$ $FeCl_3$	$CoCl_2$ $CoCl_3$	$NiCl_2$	$CuCl$ $CuCl_2$	$ZnCl_2$
YCl_3	$ZnCl_2$ $ZrCl_3$ $ZrCl_4$	$NbCl_3$ $NbCl_4$ $NbCl_5$	$MoCl_2$ $MoCl_3$ $MoCl_4$ $MoCl_5$	$TcCl_4$ $TcCl_6$	$RuCl_3$ $RuCl_4$	$RhCl_3$	$PdCl_2$	$AgCl$	$CdCl_2$
$LaCl_3$	$HfCl_3$ $HfCl_4$	$TaCl_2$ $TaCl_3$ $TaCl_4$ $TaCl_5$	WCl_2 WCl_3 WCl_4 WCl_5 WCl_6	$ReCl_3$ $ReCl_4$ $ReCl_5$	$OsCl_2$ $OsCl_3$ $OsCl_4$ $OsCl_5$	$IrCl_3$	$PtCl_2$ $PtCl_4$	$AuCl$ $AuCl_3$	Hg_2Cl_2 Hg_2Cl_2

TABLE 10

Highest oxidation state fluorides of the transition elements

Gp IIIA	Gp IVA	GP VA	Gp VIA	Gp VIIA	Gp VIII	Gp VIII	Gp VIII	Gp IB	Gp IIB
ScF_3	TiF_4	VF_5	CrF_6	MnF_4	FeF_3	CoF_3	NiF_2	CuF_2	ZnF_2
YF_3	ZrF_4	NbF_5	MoF_6	TcF_6	RuF_6	RhF_6	PdF_4	AgF_2	CdF_2
LaF_3	HfF_4	TaF_5	WF_6	ReF_7	OsF_7	IrF_6	PtF_6	AuF_5	HgF_2

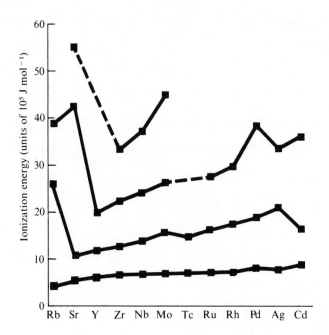

FIG. 29. Consecutive ionization energies of the second transition-series elements (compare with Fig. 25, p. 70).

chromium in chromium trioxide, CrO_3, and the chromate ion, CrO_4^{2-}, and by manganese in the permanganate ion, MnO_4^-, which as potassium permanganate is an important oxidizing agent. In the third transition series the highest group oxidation state is found for ruthenium and osmium which form the tetroxides RuO_4 and OsO_4, derived from Ru(VIII) and Os(VIII) respectively.

Table 11 summarizes the wide range of positive oxidation states exhibited by the transition metals.

Ionic and covalent bonding

The nature of the bonding in the transition-metal halides and oxides is determined by the same principles which we saw governing that in the corresponding compounds of the main-group elements. For instance we

Table 11

Positive oxidation states of the transition metals (The most stable oxidation state is circled.)

Positive oxidation state	First row							
	Ti	V	Cr	Mn	Fe	Co	Ni	Cu
8+								
7+				×				
6+			×	×	×			
5+		×	×	×		×		
4+	⊗	⊗	×	×	×	×	×	
3+	×	×	⊗	×	⊗	×	×	×
2+	×	×	×	⊗	×	⊗	⊗	⊗
1+		×	×	×	×	×	×	×

	Second row							
	Zr	Nb	Mo	Tc	Ru	Rh	Pd	Ag
8+					×			
7+				×	×			
6+			⊗	⊗	×	×		
5+		⊗	×	×	×	×		
4+	⊗	×	×	×	×	×	×	
3+	×	×	×	×	⊗	⊗		×
2+	×	×	×	×	×	×	⊗	×
1+	×	×	×	×	×	×	×	⊗

	Third row							
	Hf	Ta	W	Re	Os	Ir	Pt	Au
8+					×			
7+				⊗	×			
6+			⊗	×	×	×	×	
5+		⊗	×	×	×	×	×	×
4+	⊗	×	×	×	⊗	⊗	⊗	
3+	×		×	×	×	×	×	⊗
2+		×	×	×	×	×	×	×
1+	×	×	×	×		×	×	×

expect and find a greater tendency to covalency when the cation carries a higher charge. In agreement with this it is found that the titanium chlorides $TiCl_2$ and $TiCl_3$ are ionic compounds with high melting points, while $TiCl_4$ is a volatile liquid with covalent titanium-chlorine bonds. There are many similar examples with other transition-metal halides.

The properties of the transition-metal oxides again depend strongly on the oxidation state of the metal. The oxides in low oxidation states are usually rather unstable, but are typically basic. For instance nickel(II) oxide dissolves readily in acids to give other nickel(II) salts. Oxides in intermediate oxidation states are often amphoteric, a typical example being chromium(III) oxide, Cr_2O_3, which gives chromium(III) salts with acids and chromites with bases. Finally in high oxidation states the oxides are acidic. For instance chromium(VI) oxide, CrO_3, dissolves in basic solution to give the chromate ion, CrO_4^{2-}, and manganese(VII) oxide, Mn_2O_7, readily gives the permanganate ion, MnO_4^-. In this sense there is some formal resemblance to the main-group analogues of these compounds. Thus in group VIB sulphur trioxide, SO_3, gives the sulphate ion, SO_4^{2-}, and in group VIIB chlorine heptoxide, Cl_2O_7, gives the perchlorate ion, ClO_4^-. However there are few chemical similarities between sulphates and chromates, or between perchlorates and permanganates.

Covalent and ionic radii

The covalent radii and ionic radii (for the dipositive cations) of the elements of the first transition series are given in Fig. 30. It can be seen that the covalent radii decrease rapidly at the start of the series, then become substantially constant, and finally begin to increase at the end of the series when the filled 3d orbitals have contracted into the electron core and so shield the outer 4s electrons more effectively from the nucleus. Changes in the ionic radii on crossing the series are much less regular, so that periodic trends in the properties of these ions are difficult to rationalize. There is, however, a steady decrease in ionic radius between Mn^{2+} and Ni^{2+} and therefore a corresponding increase in the polarizing power of these ions. In turn there is a steady increase in the importance of covalency between the compounds of manganese(II) and nickel(II), and the stability of the hydrated ion $M(OH_2)_6^{2+}$ follows the series $Mn^{2+} < Fe^{2+} < Ni^{2+} < Cu^{2+} > Zn^{2+}$ in accordance (with the exception of Cu^{2+}) with the predicted polarizing powers of these ions.

Another useful comparison can be made between the covalent radii of the first and subsequent transition-series elements. These are given in Table 12.

It can be seen that there is the expected increase in radius on passing from the first to the second transition series, but that, within a group, the

Fig. 30. Covalent radii and ionic radii (for M^{2+}) of first transition-series elements.

values for the second and third transition-series elements are almost the same. The reason for this is that the lanthanide elements, in which the 4f electron shell is filled, precede the elements of the third transition series. These 4f electrons shield outer electrons inefficiently and the resulting increased effective nuclear charge causes a general contraction of the orbitals, so that the radii of the elements of the third transition series are somewhat lower than would be expected by analogy with the behaviour of main-group elements. The contraction is such that the covalent radii (and ionic radii) for the third transition series are very close to those for the second transition series. Although the elements of the second and third transition series gives larger ions, covalency is more important for them than for the elements of the first transition series. For instance while nickel(II) chloride, $NiCl_2$, is an essentially ionic compound, palladium(II) and platinum(II) chlorides exist as covalent, chlorine-bridged polymers.

The reasons for this effect are rather complex. One reason is that the first ionization energies and the electron affinities for the second and

TABLE 12

Covalent radii of the transition elements (units of nm)

Gp IIA	Gp IIIA	Gp IVA	Gp VA	Gp VIA	Gp VIIA	Gp VIII	Gp VIII	Gp VIII	Gp IB	Gp IIB
Ca 0·174	Sc 0·144	Ti 0·132	V 0·122	Cr 0·118	Mn 0·117	Fe 0·117	Co 0·116	Ni 0·115	Cu 0·117	Zn 0·125
Sr 0·191	Y 0·162	Zr 0·145	Nb 0·134	Mo 0·130	Tc 0·127	Ru 0·125	Rh 0·125	Pd 0·128	Ag 0·134	Cd 0·148
Ba 0·198	La 0·169	Hf 0·144	Ta 0·134	W 0·130	Re 0·128	Os 0·126	Ir 0·126	Pt 0·130	Au 0·134	Hg 0·149

third transition series are often higher than for the first series so that, in contrast to the usual main-group trends, the larger element atoms are the more electronegative and so form predominantly covalent compounds. In the cations of the first transition series the 3d orbitals contract into the electron core to a greater extent and are therefore less readily available to form covalent than the 4d or 5d orbitals of the second and third transition series.

Hard and soft metal ions

One of the most widely used definitions of acid–base behaviour is that of G. N. Lewis. According to this definition an acid is an electron-pair acceptor and a base is an electron-pair donor. Metal ions therefore can be considered as acids and in the example shown below the silver ion acts as an acid accepting a lone pair of electrons from the base, NH_3.

$$Ag^+ + 2NH_3 \rightarrow [Ag(NH_3)_2]^+$$

Most metal ions have a greater affinity for small, non-polarizable anions (such as O^{2-} or F^-, which can be considered as 'hard' bases) than for large, polarizable anions (such as S^{2-} or I^-, which can be considered as 'soft' bases). Such metal ions are called *hard* acids. However, some metal ions have a greater affinity for *soft* bases than for *hard* bases, as defined above, and these are called *soft* acids or *soft* metal ions. It is then convenient to remember that hard acids prefer hard bases, while soft acids prefer soft bases. Of course, there will be some metal ions which do not discriminate between the hard and soft bases, and these will have borderline hard-soft character. A classification of metals according to their hard-soft character is given in Table 13. A further complicating factor is that the hard-soft character depends on the oxidation state of an element. For example, Cu^+ is a typical soft metal ion whereas Cu^{2+} displays borderline hard-soft character. In general, an increase in oxidation state leads to an increase in the hardness of a metal ion. We have already seen that elements in high oxidation states are often combined with F^- or O^{2-}, as in PtF_6 or AgO.

Since the soft metal ions have little affinity for oxygen donors (such as H_2O), their halides and sulphides are insoluble in water. For instance nickel(II) chloride (hard) dissolves readily in water giving the $[Ni(OH_2)_6]^{2+}$ ion, but palladium(II) and platinum(II) chlorides (soft) are insoluble. Advantage may be taken of this property in quantitative or qualitative analysis. Thus the insolubility of the silver halides is used in quantitative analysis for silver or for halide ions, while the insolubility of mercury(II) or lead(II) sulphide forms the basis for qualitative tests for these metals.

TABLE 13

The 'hard' and 'soft' metal ions. Certain elements on the borderline, for example, Tc and Re, may show either 'hard' and 'soft' properties

IA	IIA	IIIA	IVA	VA	VIA	VIIA	VIII	VIII	VIII	IB	IIB	IIIB	IVB	VB
Li	Be											B	C	N
Na	Mg											Al	Si	P
K	Ca	Sc	Ti	V	Cr	Mn	Fe	Co	Ni	Cu	Zn	Ga	Ge	As
Rb	Sr	Y	Zr	Nb	Mo	Tc	Ru	Rh	Pd	Ag	Cd	In	Sn	Sb
Cs	Ba	La	Hf	Ta	W	Re	Os	Ir	Pt	Au	Hg	Tl	Pb	Bi

Hard

Soft

Transition-metal complexes

Before discussing the colours and magnetic properties of transition-metal compounds, it will be necessary to survey an important class of compounds (known as *complexes*) formed by the transition elements.

Complexes of main group elements are usually called addition compounds. Thus boron trifluoride adds ammonia according to the equation

$$H_3N: + BF_3 \rightarrow H_3N{\rightarrow}BF_3.$$

The B–N bond is an electron-pair bond in which both electrons are supplied by the nitrogen atom; such a bond is commonly called a *coordinate bond* and the product may be called a *coordination complex*.

The boron trifluoride ammonia complex is neutral, but positively or negatively charged *complex ions* may also be formed. Thus the ions $SnCl_6^{2-}$ or $Al(OH_2)_6^{3+}$ may be considered to be complex ions.

The transition elements form an especially important variety of complex compounds and complex ions, in which the co-ordination number six is very common. For instance chromium(III) chloride dissolves in water to give the complex ion $Cr(OH_2)_6^{3+}$, from which the water molecules can be displaced by ammonia according to the equation

$$Cr(OH_2)_6^{3+} + 6NH_3 \rightarrow Cr(NH_3)_6^{3+} + 6H_2O.$$

In transition-metal complexes the groups attached to the metal are called *ligands* so that the ions $Cr(OH_2)_6^{3+}$ or $Cr(NH_3)_6^{3+}$ contain the ligands water or ammonia respectively. Ligands can be either neutral or negatively charged; thus the complex ion $PtCl_6^{2-}$ would be derived from the Pt^{4+} cation surrounded by six Cl^- ligands.

In these six coordinate complexes the ligands are almost always at the corners of an octahedron with the metal atom at the centre. The $Cr(NH_3)_6^{3+}$ ion is represented below

$$\left[\begin{array}{ccc} & NH_3 & \\ H_3N \cdots & | & \cdots NH_3 \\ & Cr & \\ H_3N & | & NH_3 \\ & NH_3 & \end{array} \right]^{3+}$$

If the attraction between the metal ion and the ligands is considered to be electrostatic in origin, the colours and magnetic properties of transition-metal compounds can be explained in terms of a simple bonding theory known as crystal-field theory.

Crystal-field theory

In Crystal-field theory, it is assumed that the only interaction between the metal centre and the ligands is an electrostatic (ionic) one. For example, we consider the complex ion TiF_6^{3-} to be formed from the cation Ti^{3+}, whose electron configuration is (Ar core) $3d^1$, by bringing up six F^- ligands along the x, y, and z axes as shown in Fig. 31.

In the free ion Ti^{3+} the odd d electron has an equal chance of being in any of the five 3d orbitals since they are equal energy (degenerate). However an examination of the representations of d orbitals in Fig. 4 shows that in the complex TiF_6^{3-} the lobes of the $d_{x^2-y^2}$ and d_{z^2} orbitals are directed *along* the axes and so directly *towards* the F^- ions while those of the d_{xy}, d_{yz}, and d_{xz} orbitals are directed *between* the axes and so between the F^- ions. The electron will prefer to occupy an orbital whose lobes project between the F^- ions since the repulsive electrostatic forces between the electron and the negatively charged ligands will then be minimized. Thus the d orbitals are split into two sets of unequal energy by

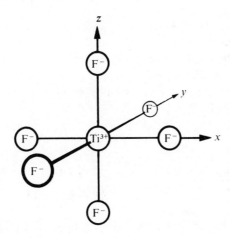

FIG. 31. Crystal-field representation of the TiF_6^{3-} ion.

the *octahedral crystal field* as shown below:

$$\text{Energy} \uparrow \quad \textcircled{1} \; \bigcirc \; \bigcirc \; \bigcirc \; \bigcirc \quad \rightarrow \quad \begin{array}{c} \bigcirc \; \bigcirc \quad \uparrow \; d_{x^2-y^2}, d_{z^2} \\[4pt] \textcircled{1} \; \bigcirc \; \bigcirc \quad \downarrow \; d_{xy}, d_{yz}, d_{xz} \end{array}$$

Degenerate 3d orbitals   Degeneracy removed by F^- ligands

The $d_{x^2-y^2}$ and d_{z^2} orbitals are relatively destabilized whilst the other d orbitals are stabilized.

The energy difference between the three lower-energy orbitals and the two upper-energy orbitals is called the *crystal-field splitting* and is given the symbol Δ.

A similar splitting of the d orbitals occurs when the ligands are neutral molecules as in the $Ti(OH_2)_6^{3+}$ ion. Here the electrostatic repulsions which give rise to the splitting are between the lone pairs of electrons on the oxygen atoms of water (p. 53) and the 3d electron on the Ti^{3+} ion.

The size of the crystal-field splitting depends on the nature of the ligands. Furthermore, the precise geometrical arrangement of the ligands around the metal ion will affect which orbitals are stabilized and which are destabilized. For example, in VCl_4 the four Cl^- ligands are arranged in a tetrahedral fashion around the $V^{4+}(d^1)$ ion. The degeneracy of the 3d orbitals is removed and the orbital splitting pattern is as shown below.

$$\textcircled{1} \; \bigcirc \; \bigcirc \; \bigcirc \; \bigcirc \quad \rightarrow \quad \begin{array}{c} \bigcirc \; \bigcirc \; \bigcirc \quad \uparrow \; d_{x^2-y^2}, d_{z^2} \\[4pt] \textcircled{1} \; \bigcirc \quad \downarrow \; d_{xy}, d_{yz}, d_{xz} \end{array}$$

Colours of transition-metal complexes

When light passes through a solution containing the $Ti(OH_2)_6^{3+}$ ion, a quantum of light may be absorbed by the ion, thus exciting an electron from a lower-level to a higher-level d orbital:

$$\begin{array}{c} \bigcirc \; \bigcirc \\ \textcircled{1} \; \bigcirc \; \bigcirc \end{array} \rightarrow \begin{array}{c} \textcircled{1} \; \bigcirc \quad d_{x^2-y^2}, d_{z^2} \\ \bigcirc \; \bigcirc \; \bigcirc \quad d_{xy}, d_{yz}, d_{xz} \end{array}$$

The frequency v of the light absorbed will be given by the expression

$$E = hv = \Delta,$$

the crystal-field splitting.

For the $Ti(OH_2)_6^{3+}$ ion, the crystal-field splitting is such that the light absorbed is in the visible part of the spectrum and the ion is coloured. In fact yellow light is absorbed and most of the blue and red light is transmitted; the result is that solutions containing the $Ti(OH_2)_6^{3+}$ ion are violet in colour.

It happens that the complexes of the first transition series usually absorb visible light and a great variety of colours are observed, the precise colour depending on the frequency of the light absorbed. Thus in aqueous solution, salts of chromium(III) are violet, those of manganese(II) are pale pink and those of copper(II) are bright blue.

In contrast, complexes of the elements of the second and third transition series are less highly coloured and are commonly pale yellow or white. For these elements the d orbitals project further into space towards the ligands so that the electrostatic repulsions between the ligands and the d electrons are increased. In turn this leads to a high crystal-field splitting, so that high-energy light (usually in the ultra-violet region of the spectrum) is absorbed, and the complexes are yellow or colourless.

Magnetic properties of transition-metal complexes

We have seen that transition-metal ions often have odd or unpaired d electrons; these impart a magnetic moment to the transition-metal ion. When a magnetic field is applied, this magnetic moment causes the transition-metal complex to be attracted into the stronger part of the field; the complex is said to be *paramagnetic*. Now the magnitude of this attractive force increases with the number of unpaired electrons on the transition-metal ion. Thus complexes of the Ti^{3+} ion having a single unpaired d electron interact less strongly with a magnetic field than complexes of V^{3+} which have two unpaired d electrons. Complexes containing no unpaired electrons are said to be *diamagnetic* and are weakly repelled by a magnetic field.

Let us consider the electronic configurations of transition-metal ions having several d electrons. For example, the Fe^{3+} ion has five 3d electrons. In octahedral complexes of this ion, two probable electron configurations are A and B

$$\textcircled{1}\ \textcircled{1} \qquad \bigcirc\bigcirc \quad d_{z^2},\ d_{x^2-y^2}$$

$$\textcircled{1}\ \textcircled{1}\ \textcircled{1} \qquad \textcircled{1\!\!\!1}\ \textcircled{1\!\!\!1}\ \textcircled{1}\ d_{xy},\ d_{yz},\ d_{xz}.$$

Configuration A   Configuration B

As noted earlier (p. 11) the electrons prefer to keep their spins parallel, and a definite energy, P, is required to pair the electron spins. Therefore if the crystal-field splitting Δ is small, the electron configuration A with five unpaired electrons will be adopted, while if Δ is greater than the pairing energy P, the electron configuration B with only one unpaired electron will be preferred.

Now in octahedral ion(III) complexes the *high-spin* electron configuration A is almost always adopted and a corresponding high

TABLE 14

Numbers of unpaired electrons and magnetic moments for some transition-metal ions

Electron configuration		Examples	Number of unpaired electrons	Magnetic moment (BM)
d^1		Ti^{3+}, V^{4+}	1	0·6–1·8
d^2		V^{3+}, Cr^{4+}	2	2·7–2·9
d^3		V^{2+}, Cr^{3+}, Mn^{4+}	3	3·7–4·0
d^4	high spin	Cr^{2+}, Mn^{3+}	4	4·7–5·0
	low spin	Cr^{2+}, Mn^{3+}	2	3·2–3·3
d^5	high spin	Mn^{2+}, Fe^{3+}	5	5·6–6·1
	low spin	Mn^{2+}, Fe^{3+}	1	1·8–2·5
d^6	high spin	Fe^{2+}	4	5·7–6·1
	low spin	Fe^{2+}, Co^{3+}	0	
d^7	high spin	Co^{2+}	3	4·3–5·2
	low spin	Co^{2+}, Ni^{3+}	1	1·8–2·0
d^8		Ni^{2+}	2	2·9–3·3
d^9		Cu^{2+}	1	1·7–2·2

paramagnetism is found. However, since the crystal-field splitting Δ is much greater for complexes of the second and third transition-series elements the *low-spin* electron configuration B is always found in octahedral complexes of ruthenium(III) and osmium(III). These complexes therefore have a low magnetic moment and are more weakly paramagnetic than the iron(III) complexes. The presence of more unpaired electrons in complexes of the first transition-series elements compared with those of the second and third transition series in a quite general phenomenon when alternative high-spin or low-spin electron configurations are possible.

Table 14 lists the number of unpaired electrons for the first row transition-metal ions, in their common oxidation states, along with experimentally determined magnetic moments, (units, Bohr magnetons, BM).

As can be seen from Table 14, measurements of the magnetic moment of a transition-metal complex can be used to estimate the number of unpaired electrons. In favourable cases, it is then possible to deduce the geometry of the complex as well as something about the nature of the metal-ligand bonding.

Summary of transition-metal properties

We have seen that the transition elements have the common property of forming compounds in several oxidation states, but that the elements of the first transition series differ in several respects from the heavier transition elements. The differences which we have discussed in this chapter may be listed as follows:

(1) The first transition-series elements form high oxidation-state complexes less readily than the heavier transition metals. In particular the oxidation state two is much more common in the first transition series than in subsequent series.

(2) The complexes of the first transition-series elements exhibit a much larger range of colours and are often more strongly paramagnetic than those of the heavier elements.

(3) Covalent bonding is less important for the first transition-series elements in low oxidation state than for the second and third series. The heavier elements are more likely to exhibit soft metal behaviour, and so form water-insoluble halides and sulphides, than the elements of the first transition series.

PROBLEMS

7.1. Why are scandium and zinc not usually treated as transition metals?

7.2. Explain why 4s rather than 3d electrons are preferentially lost on forming transition metal *cations*, while the 4s are filled before the 3d orbitals during the building up of the periodic table to form the neutral atoms.

7.3. The ion Co^{3+} has the electron configuration (Ar core) $3d^6$ and has four unpaired electrons, yet there are no unpaired electrons in the complex ion $Co(NH_3)_6^{3+}$. How can you explain this? (Hint: see p. 8).

7.4. Which of the following pairs of compounds do you expect to be the more soluble in water? Why?
    (a) LiF, LiI   (b) AgF, AgI

7.5. Explain why the $d_{x^2-y^2}$ and d_{z^2} orbitals are stabilized and the d_{xy}, d_{yz}, d_{xz} orbitals are destabilized in the tetrahedral splitting pattern of page 86? (Hint: consider shape of d orbitals.)

8. Periodicity and chemical compounds: the lanthanides and actinides

The lanthanides and actinides are the elements in the periodic table which are formed as the 4f or 5f orbitals are progressively filled. There are therefore fourteen elements in each series, and all are metallic in nature.

The lanthanide elements and their ground-state electron configurations are listed in Table 15. The rather irregular trends in the electron configurations of these elements indicate that the valence orbitals (the 4f, 5d, and 6s orbitals) are very close to each other in energy. Table 15 also provides evidence for the increased stability of electron configurations having half-filled ($4f^7$) or filled ($4f^{14}$) shells of f electrons.

On forming positively charged ions, the 4f orbitals are strongly stabilized with respect to the other valence orbitals (5d and 6s). In fact this stabilization of the 4f orbitals in the positively charged lanthanide elements is so great that, by the time the tripositive lanthanide ions have been formed, the 4f electrons have entered the electron core, with the

TABLE 15

Ground-state electron configurations of the lanthanide elements

Element	Symbol	Ground state		
		4f	5d	6s
Lanthanum	La	0	1	2
Cerium	Ce	1	1	2
Praseodymium	Pr	3		2
Neodymium	Nd	4		2
Promethium	Pm	5		2
Samarium	Sm	6		2
Europium	Eu	7		2
Gadolinium	Gd	7	1	2
Terbium	Tb	9		2
Dysprosium	Dy	10		2
Holmium	Ho	11		2
Erbium	Er	12		2
Thulium	Tm	13		2
Ytterbium	Yb	14		2
Lutetium	Lu	14	1	2

result that the chemistry of the lanthanide elements is dominated by the formation of M^{3+} ions. In the actinide elements the energies of the 5f orbitals are less sensitive to increases in nuclear charge on forming positively charged ions, so that oxidation states higher than three can be reached more readily by these elements than by the lanthanides.

After this brief introduction, the chemistry of the lanthanide elements will be discussed and this will then be compared with the rather different chemistry of the actinide elements.

Binding energies of the lanthanide elements

The binding energies and melting points for the lanthanide elements are given in Fig. 32. It can be seen that the binding energies are almost constant on crossing the series; the magnitudes indicate that for all the elements there are just three electrons which contribute to the binding. Bearing in mind the properties of the transition elements (p. 72), it is not

FIG. 32. Binding energies (●——) and melting points (▲---) of the lanthanide elements.

surprising that each lanthanide element also uses just three valence electrons in chemical combination with other elements.

Oxidation states of the lanthanides

The most stable oxidation state for all the lanthanide elements is three. Thus they all form ionic fluorides of general formula MF_3 and basic oxides M_2O_3.

For most of the lanthanides the *only* oxidation state is three, but there are certain exceptions. Europium and ytterbium can also form dipositive ions, Eu^{2+} and Yb^{2+}, which have the respective electron configurations (Xe core) $4f^7$ and (Xe core)$4f^{14}$ with favourable half-filled or filled shells of 4f electrons. Similarly cerium and terbium can form compounds in oxidation state four; the respective electron configurations for Ce^{4+} and Tb^{4+} are (Xe core) $4f^0$ and (Xe core) $4f^7$ respectively. Even in these cases however, the oxidation state three is the more stable. Thus compounds of Eu(II) are easily oxidized to Eu(III), while Ce(IV) compounds are easily reduced to Ce(III).

Covalent and ionic radii of the lanthanides

The covalent radii for the lanthanide elements and the radii of their tripositive ions are shown in Fig. 33. In each case there is a general decrease in the radii on passing across the lanthanide series due to the general increase in effective nuclear charge as the 4f electron shell is filled. This arises because the 4f electrons shield each other from the nuclear charge rather poorly, so that as more 4f electrons are added the effective nuclear charge felt by each of them increases. The result is that the whole 4f electron shell contracts on passing across the lanthanide series. This steady decrease in covalent and ionic radii on crossing the series is generally called the *lanthanide contraction*.

The very large irregularity seen in the covalent radius of europium, Eu, is explained in terms of the lattice structure of this element. The other lanthanide elements are composed of a lattice of Ln^{3+} ions with a $4f^n$ configuration and three electrons in the valence band. Metallic europium, however, is composed predominantly of Eu^{2+} ions, having the stable $4f^7$ configuration and there are only two electrons in the valence band. The effective nuclear charge felt by the outer electrons will be less than for Ln^{3+}, producing the very large covalent radius.

We have already seen (p. 79) that the lanthanide contraction is responsible for the similarity in covalent and ionic radii between the elements of the second and third transition series. The effect is also important in the chemistry of the lanthanides themselves, since the general decrease in ionic radii on crossing the series leads to a corresponding increase in the polarizing power of the ions and in the stability of complexes of the ions.

FIG. 33. Covalent radii (▲——) and ionic radii for M^{3+} ions (●——) of the lanthanides.

The actinide elements

Prior to 1940 only the naturally occurring actinides, thorium, protoactinium, and uranium, were known. These elements beyond actinium (89) were believed to be the first three of a fourth transition series, and when the elements beyond uranium were discovered, they were at first called the *transuranic elements* and put into the main body of the Periodic Table (see front cover): for example, neptunium (93) was placed under Group VIIA, uranium (92) being in Group VIA. When the remainder of the elements (94–103) were discovered, it became clear that they formed a series analogous to the lanthanides, and they were called *actinides*, and grouped in the Periodic Table with the lanthanides. Elements 94–103 were produced artificially. For example, neptunium was synthesized by neutron bombardment as shown below:

$$^{238}_{92}U + ^{1}_{0}n \rightarrow ^{239}_{92}U \rightarrow ^{239}_{93}Np + \beta^{-}.$$

Element 103, lawrencium, was synthesised in 1961, also by means of a nuclear bombardment reaction.

The actinide elements are all radioactive and this has prevented a

TABLE 16

Ground-state electron configurations of the actinide elements

Element	Symbol	Ground state		
		5f	6d	7s
Actinium	Ac		1	2
Thorium	Th		2	2
Protoactinium	Pa	2	1	2
	or	1	2	2
Uranium	U	3	1	2
Neptunium	Np	5		2
Plutonium	Pu	6		2
Americium	Am	7		2
Curium	Cm	7	1	2
Berkelium	Bk	8	1	2
	or	9		2
Californium	Cf	10		2
Einsteinium	Es	11		2
Fermium	Fm	12		2
Mendelevium	Md	13		2
Nobelium	No	14		2
Lawrencium	Lr	14	1	2

thorough investigation of the properties of many of them. The elements and their probable ground-state electron configurations are given in Table 16.

As might be expected, the actinides have much in common with the lanthanides, but there are also important differences. These arise because the 5f orbitals of the actinide elements do not drop in energy to such a great extent as do the 4f orbitals of the lanthanides when positive ions are formed. For this reason, the actinide elements often form compounds in which their oxidation states are higher than three.

Oxidation states of the actinide elements

The important oxidation states of the actinide elements are shown in Table 17. It can be seen that, like the lanthanides, the actinides all form compounds in oxidation state three, but that higher oxidation states are possible and indeed are favoured at the start of the series. Thus thorium(IV), protoactinium(V), and uranium(VI) compounds are all more stable than the compounds of these elements in oxidation state

Table 17

Oxidation states of the actinide elements. The most important oxidation state(s) for each element are underlined

Ac	Th	Pa	U	Np	Pu	Am	Cm	Bk	Cf	Es	Fm	Md	No
_3_	3	3	3	3	3	_3_	_3_	_3_	_3_	_3_	_3_	_3_	_3_
	4		_4_	_4_	_4_	4	4	4					
		_5_	5	_5_	5	5							
			_6_	_6_	6	6							

three. At the start of the actinide series then, the behaviour is more reminiscent of the transition elements than of the lanthanides, but later in the series the oxidation state three becomes progressively more important until at the end of the actinide series the properties of the elements and their compounds are very similar to those of the lanthanides.

Other similarities with the lanthanides can be attributed more directly to the presence of partially-filled shells of f electrons. Thus, like the transition-element complexes, the compounds of both the lanthanides and actinides are often coloured and paramagnetic.

Recently discovered elements

Elements 104 and 105 have been reported independently, in recent times, by American and Soviet workers. Due to some disagreement over the naming of these new elements, the International Union of Pure and Applied Chemistry (IUPAC) has recommended a systematic name for these elements. This nomenclature is based on a mixture of Greek and Latin roots, which has been found necessary to avoid ambiguities at this point. Thus element 104 is loosely named after the Greek for 104, unnilquadium, (Unq), and element 105 is named unnilpentium, (Unp). These same two research groups have synthesized element 106, unnilhexium, (Unh) and evidence also exists for element 107, unnilseptium, (Uns). Clearly, there is still room for further expansion of the periodic table, as the discovery of further new elements is predicted. Further, the chemical properties of such new elements can be predicted by use of the periodic table, just as Mendeleev predicted the properties of then undiscovered elements over one hundred years ago.

PROBLEMS

8.1. The elements with atomic number 104–107 have recently been detected. Can you predict any of their chemical properties from their expected position in the periodic table?

Bibliography and notes

Greenwood, N. N. and Earnshaw, A. *Chemistry of the elements*. Pergamon Press, Oxford, (1984). This is a comprehensive textbook on inorganic chemistry, giving a detailed account of chemical bonding and a thorough treatment of individual elements. It serves as an excellent reference book for senior university students.

Cotton, F. A. and Wilkinson, G. *Advanced inorganic chemistry*, 4th ed. Interscience Publishers, New York, (1980). This textbook has, for many years, been a leading influence on the development of inorganic chemistry. The content is of an advanced nature and the book is mainly suitable for senior university students and beyond.

Huheey, J. E. *Inorganic chemistry*, 3rd ed. Harper and Row Publishers, New York (1983). This is a very well written book. Its emphasis is on structural and bonding principles with less space given to descriptive chemistry.

Douglas, B. E., McDaniel, D. H. and Alexander, J. J. *Inorganic chemistry*, 2nd ed. John Wiley and Sons, Inc., New York (1983). This book presents the principles of inorganic chemistry in a very readable fashion. The emphasis is on bonding and structure rather than a detailed study of individual elements.

Coulson, C. A. revised McWeeny, R. *The shape and structure of molecules*, 2nd ed. Clarendon Press, Oxford (1982). This is an excellent introduction to the important topic of chemical bonding. The book touches on all the important concepts and words and avoids any involved mathematical treatment of bonding. It is suitable for first and second year university students.

Answers to problems

2.1. (a) $n = 2, l = 0, m = 0,$ $\qquad\qquad s = +\frac{1}{2}$ or $-\frac{1}{2}$
(b) $n = 2, l = 1, m = 1, 0,$ or $-1,$ $\qquad s = +\frac{1}{2}$ or $-\frac{1}{2}$
(c) $n = 3, l = 2, m = 2, 1, 0, -,$ or $-2, s = +\frac{1}{2}$ or $-\frac{1}{2}$

Two electrons can be accommodated in the 2s orbital, six electrons in the 2p orbitals, and ten electrons in the 3d orbitals.

3.1. Atomic number	Electron configuration	Element	Group	Block
4	$1s^2 2s^2$	Be	II	s-block
16	$1s^2 2s^2 2p^6 3s^2 3p^4$	S	VI	p-block
22	$1s^2 2s^2 2p^6 3s^2 3p^6 3d^2 4s^2$	Ti	IVA	d-block

4.1. Ionization energy ⎱ ⎰(a) B < N < F < Ne
Electronegativity ⎰ ⎱(b) Ba < Sr < Ca < Mg < Hg

4.2. $Na^+ > Li^+ > Be^{2+} > B^{3+}$. B^{3+} is formed least readily.

4.3. Compound	Oxidation state	Co-ordination number
$SiCl_4$	4	4
$SiCl_6^{2-}$	4	6
$AlCl_4^-$	3	4
AlF_6^{3-}	3	6

5.2. (a) Si > Al > Mg > Na > K. This is the order of decreasing binding energies.
(b) S > P > Cl > Ar. This is the order of decreasing molecular size; i.e. S_8 > P_4 > Cl_2 > Ar.

5.3 (a) Al > Si > P
(b) Po > Te > Se

6.2. BCl_3, $SiCl_4$, and PCl are hydrolysed rapidly by water

7.3. In $Co(NH_3)_3^{3+}$, cobalt has the electron configuration

with no unpaired electrons, while the free ion Co^{3+} has the configuration

with four unpaired electrons.

7.4. Solubility of LiI > LiF but AgF > AgI. Lithium is a hard metal ion, while silver is a soft metal ion.

8.1. Element 104 falls into group IVA and should resemble Ti, Zr, and Hf in its chemical properties.

Index

SI units

Physical quantity	Old unit	Value in SI units
energy	calorie (thermochemical)	4·184 J (joule)
	*electronvolt—eV	$1·602 \times 10^{-19}$ J
	*electronvolt per molecule	96·48 kJ mol^{-1}
	erg	10^{-7} J
	*wave number—cm^{-1}	$1·986 \times 10^{-23}$ J
entropy (S)	eu = cal g^{-1} °C^{-1}	4184 J kg^{-1} K^{-1}
force	dyne	10^{-5} N (newton)
pressure (P)	atmosphere	$1·013 \times 10^5$ Pa (pascal), or N m^{-2}
	torr = mmHg	133·3 Pa
dipole moment (μ)	debye—D	$3·334 \times 10^{-30}$ C m
magnetic flux density (H)	*gauss—G	10^{-4} T (tesla)
frequency (ν)	cycle per second	1 Hz (hertz)
relative permittivity (ε)	dielectric constant	1
temperature (T)	*°C and °K	1 K (kelvin); 0 °C = 273·2 K

(* indicates permitted non-SI unit)

Multiples of the base units are illustrated by length

fraction	10^9	10^6	10^3	1	(10^{-2})	10^{-3}	10^{-6}	10^{-9}	(10^{-10})	10^{-12}
prefix	giga-	mega-	kilo-	metre	(centi-)	milli-	micro-	nano-	(*ångstrom)	pico-
unit	Gm	Mm	km	m	(cm)	mm	μm	nm	(*Å)	pm

The fundamental constants

Avogadro constant	L or N_A	$6·022 \times 10^{23}$ mol^{-1}
Bohr magneton	μ_B	$9·274 \times 10^{-24}$ J T^{-1}
Bohr radius	a_0	$5·292 \times 10^{-11}$ m
Boltzmann constant	k	$1·381 \times 10^{-23}$ J K^{-1}
charge of a proton	e	$1·602 \times 10^{-19}$ C
(charge of an electron = $-e$)		
Faraday constant	F	$9·649 \times 10^4$ C mol^{-1}
gas constant	R	$8·314$ J K^{-1} mol^{-1}
nuclear magneton	μ_N	$5·051 \times 10^{-27}$ J T^{-1}
permeability of a vacuum	μ_0	$4\pi \times 10^{-7}$ H m^{-1} or N A^{-2}
permittivity of a vacuum	ε_0	$8·854 \times 10^{-12}$ F m^{-1}
Planck constant	h	$6·626 \times 10^{-34}$ J s
(Planck constant)/2π	$\hbar$	$1·055 \times 10^{-34}$ J s
rest mass of electron	m_e	$9·110 \times 10^{-31}$ kg
rest mass of proton	m_p	$1·673 \times 10^{-27}$ kg
speed of light in a vacuum	c	$2·998 \times 10^8$ m s^{-1}

$\ln 10 = 2·303$ $\ln x = 2·303 \lg x$ $\lg e = 0·4343$ $\pi = 3·142$
$R \ln 10 = 19·14$ J K^{-1} mol^{-1} $RTF^{-1} \ln 10 = 59·16$ mV at 298·2 K

Periodic Table

New IUPAC notation (top numbers) / Previous IUPAC form (letter groups)

1	2	3	4	5	6	7	8	9	10	11	12	13	14	15	16	17	18
IA	IIA	IIIA	IVA	VA	VIA	VIIA	VIII	VIII	VIII	IB	IIB	IIIB	IVB	VB	VIB	VIIB	O
1 H 1·008																	2 He 4·003
3 Li 6·941	4 Be 9·012											5 B 10·81	6 C 12·01	7 N 14·01	8 O 16·00	9 F 19·00	10 Ne 20·18
11 Na 22·99	12 Mg 24·31											13 Al 26·98	Si 28·09	15 P 30·97	16 S 32·45	17 Cl 35·45	18 A 39·95
19 K 39·10	20 Ca 40·08	21 Sc 44·96	22 Ti 47·90	23 V 50·94	24 Cr 52·00	25 Mn 54·94	26 Fe 55·85	27 Co 58·93	28 Ni 58·71	29 Cu 63·55	30 Zn 65·37	31 Ga 69·72	32 Ge 72·92	33 As 74·92	34 Se 78·96	35 Br 79·90	36 Kr 83·80
37 Rb 85·47	38 Sr 87·62	39 Y 88·91	40 Zr 91·22	41 Nb 92·91	42 Mo 95·94	43 Tc 98·91	44 Ru 101·1	45 Rh 102·9	46 Pd 106·4	47 Ag 107·9	48 Cd 112·4	49 In 114·8	50 Sn 118·7	51 Sb 121·8	52 Te 127·6	53 I 126·9	54 Xe 131·3
55 Cs 132·9	56 Ba 137·3	57 La 138·9	72 Hf 178·5	73 Ta 180·9	74 W 183·9	75 Re 186·2	76 Os 190·2	77 Ir 192·2	78 Pt 195·1	79 Au 197·0	80 Hg 200·6	81 Tl 204·4	82 Pb 207·2	83 Bi 209·0	84 Po (210)	85 At (210)	86 Rn (222)
87 Fr (223)	88 Ra 226·0	89 Ac (227)	104 Unq	105 Unp	106 Unh	107 Uns											

← s – block →   ← d – block →   ← p – block →

Lanthanides	57 La 138·9	58 Ce 140·1	59 Pr 140·9	60 Nd 144·2	61 Pm (147)	62 Sm 150·4	63 Eu 152·0	64 Gd 157·3	65 Tb 158·9	66 Dy 162·5	67 Ho 164·9	68 Er 167·3	69 Tm 168·9	70 Yb 173·0	71 Lu 175·0
Actinides	89 Ac (227)	90 Th 232·0	91 Pa 231·0	92 U 238·0	93 Np 237·0	94 Pu (242)	95 Am (243)	96 Cm (248)	97 Bk (247)	98 Cf (251)	99 Es (254)	100 Fm (253)	101 Md (256)	102 No (254)	103 Lr (257)

f – block